ALGORITHMS
FOR
HVAC ACOUSTICS

About the Authors

This publication was completed as part of the American Society of Heating, Refrigerating, and Air-Conditioning Engineers, Inc., Research Project 556-RP, Algorithms for HVAC Acoustics, by DDR, Inc., Consultants in Acoustics & Vibration, Las Vegas, NV.

Douglas D. Reynolds, Ph.D., is a Professor of Mechanical Engineering in the Department of Mechanical Engineering at the University of Nevada, Las Vegas. He is also President of D.D.R., Inc.

Jeffrey M. Bledsoe was a Graduate Research Assistant in the Department of Civil and Mechanical Engineering at the University of Nevada, Las Vegas. Currently he is employed by the Ford Motor Company as a Test Engineer at its Arizona Proving Grounds.

ASHRAE Special Publications

W. Stephen Comstock
 Director, Communications/Publications
Mildred Geshwiler
 Editor, Special Publications
Lynn Montgomery
 Associate Editor
Michelle Moran
 Associate Editor
Jacquelyn Johnson
 Editorial Secretary

ALGORITHMS FOR HVAC ACOUSTICS

Douglas D. Reynolds
Jeffrey M. Bledsoe

American Society of Heating, Refrigerating
and Air-Conditioning Engineers, Inc.

ISBN 0-910110-75-1

Copyright 1991
American Society of Heating, Refrigerating,
and Air-Conditioning Engineers, Inc.
1791 Tullie Circle N.E.
Atlanta, Georgia 30329

§ § §

No part of this book may be reproduced without permission in writing from ASHRAE, except by a reviewer who may quote brief passages or reproduce illustrations in a review with appropriate credit; nor may any part of this book be reproduced, stored in a retrieval system, or transmitted in any form or by any means (electronic, photocopying, recording, or other) without permission in writing from ASHRAE.

The computer programs in this publication are presented under the terms of a license agreement between the American Society of Heating, Refrigerating, and Air-Conditioning Engineers, Inc., and D.D.R., Inc.

The computer programs presented in this publication may be used, either individually or grouped together to form an integrated computer program, for personal or internal uses only. The computer programs presented in this publication, either individually or grouped together to form an integrated computer program, may not be published, distributed, or transmitted in any form or by any means (including but not limited to written form, photocopying, recording, electronic, magnetic, optical, or other); nor may the computer programs be stored in a retrieval system without permission in writing from D.D.R., Inc.

Even though great care has been exercised in the preparation and writing of this publication and of the computer programs contained in this publication, ASHRAE and D.D.R., Inc., assume no responsibility for errors or omissions; nor is any liability assumed for damages resulting from the use of the information or computer programs contained herein.

§ § §

The views of the authors do not purport to reflect the position of the Society, nor does their publication imply any recommendation or endorsement thereof by the Society.

TABLE OF CONTENTS

1. **INTRODUCTION**	1–1
2. **SOME BASICS**	2–1
2.1. Sound Levels	2–1
2.2. Noise Criterion Curves	2–4
2.3. Room Criterion Curves	2–6
3. **EQUIPMENT SOUND POWER**	3–1
3.1. Fans	3–1
3.2. Chillers	3–2
4. **DUCT ELEMENT REGENERATED SOUND POWER**	4–1
4.1. Dampers	4–1
4.2. Elbows Fitted with Turning Vanes	4–3
4.3. Junctions and Turns	4–5
4.4. Diffusers	4–12
5. **DUCT ELEMENT SOUND ATTENUATION**	5–1
5.1. Sound Plenums	5–2
5.2. Unlined Rectangular Sheet Metal Ducts	5–4
5.3. Acoustically Lined Rectangular Sheet Metal Ducts	5–5
5.4. Unlined Circular Sheet Metal Ducts	5–7
5.5. Acoustically Lined Circular Sheet Metal Ducts	5–7
5.6. Rectangular Sheet Metal Elbows	5–9
5.7. Acoustically Lined Circular Radiused Elbows	5–12
5.8. Duct Silencers	5–13
5.9. Duct Branch Power Division	5–20
5.10. Duct End Reflection Loss	5–21
5.11. Terminal Volume Regulation Units	5–22
6. **DUCT BREAKOUT AND BREAKIN**	6–1
6.1. Sound Breakout and Breakin	6–1
6.2. Rectangular Ducts	6–2
6.3. Circular Ducts	6–5
6.4. Flat Oval Ducts	6–8
6.5. Insertion Loss of External Acoustic Lagging on Rectangular Ducts	6–11
7. **SOUND TRANSMISSION IN INDOOR AND OUTDOOR SPACES**	7–1
7.1. Sound Attenuation through Ceiling Systems	7–1
7.2. Receiver Room Sound Corrections	7–2
7.3. Sound Transmission through Mechanical Equipment Room Walls, Floor or Ceiling	7–3
7.4. Sound Transmission in Outdoor Environments	7–11
8. **GENERAL INFORMATION ON THE DESIGN OF HVAC SYSTEMS**	8–1
9. **CONCLUSIONS AND RECOMMENDATIONS**	9–1

REFERENCES

APPENDIX 1:   Computer Programs for Chapter 2

APPENDIX 2:   Computer Programs for Chapter 3

APPENDIX 3:   Computer Programs for Chapter 4

APPENDIX 4:   Computer Programs for Chapter 5

APPENDIX 5:   Computer Programs for Chapter 6

APPENDIX 6:   Computer Programs for Chapter 7

INTRODUCTION

CHAPTER 1

Over the past 10 years ASHRAE Technical Committee TC 2.6, Sound and Vibration Control, has sponsored research that has greatly expanded the available technical data associated with HVAC acoustics. These data, all of which have been included in the ASHRAE Handbook chapter on sound and vibration control, have greatly expanded the ability of designers to make more accurate calculations related to the acoustical characteristics of HVAC systems. However, one of the major difficulties associated with HVAC analyses has been the enormous amount of time that is required to make the many repetitive calculations necessary to examine the sound properties of a specific system. In order for individuals to analyze the sound characteristics of a particular system and to make recommendations that are necessary to ensure a quiet system, the technical data and associated design procedures developed under the sponsorship of TC 2.6 must be presented in a format that is easy to use. As a result, ASHRAE Research Project 556-RP, Algorithms for HVAC Acoustics, was approved for the purpose of developing for the HVAC designer a set of algorithms and related computer programs in the area of HVAC acoustics that are useful and reliable. All algorithms were to be based on currently verifiable published and unpublished test results.

The objectives of this project were to: (1) develop algorithms in English, along with discussions related to the development of the algorithms, references, and other appropriate data; (2) develop computer programs associated with each algorithm programmed in Basic; and (3) produce a final report in a form that can be readily published by ASHRAE.

Algorithms that are presented in this report fall under the following general headings:

Some Basics
Equipment Sound Power
Duct Element Regenerated Sound Power
Duct Element Sound Attenuation
Duct Breakout and Breakin
Sound Transmission in Indoor and Outdoor Spaces

The section on "Some Basics" covers basic terminology, the addition of sound levels, and the determination of NC and RC levels.

The algorithms developed under the general heading of "Equipment Sound Power" are "Fans" and "Chillers." The algorithm for fans is based on the methodology presented in the ASHRAE Handbook for determining the 1/1 octave band sound power levels of fans. The algorithm for chillers is also based on the ASHRAE Handbook. Methods presented here for determining sound pressure levels for chillers should be used for approximations only. Noise problems arising from water-chilling assemblies are dependent upon so many variables that it is impossible to formulate a standard solution. Chiller manufacturers have yet to publish any substantial data due to a lack of an industry-sanctioned, standard method of determining and reporting noise levels.

The algorithms developed under the general heading of "Duct Element Regenerated Sound Power" include: dampers, elbows fitted with turning vanes, junctions and turns, and diffusers. Algorithms for dampers, elbows fitted with turning vanes, and junctions and turns are based on ASHRAE-sponsored research performed by Ver and information in the ASHRAE Handbook (36,44,45,46,47). The algorithm for diffusers is based on material presented in Beranek's book, Noise and Vibration Control (3).

Algorithms developed under the general heading of "Duct Element Sound Attenuation" include: sound plenums, unlined rectangular ducts, acoustically lined rectangular ducts, unlined circular ducts, acoustically lined circular ducts, elbows, acoustically lined circular radiused elbows, duct silencers, duct branch power division, duct end reflection loss, and terminal volume regulation units. The algorithm for plenum chambers is based on work originally completed by Wells (48). Very little research has been undertaken investigating the acoustical properties of plenum chambers since Wells published his paper; thus, further experimental work needs to be performed in this area. Algorithms for unlined rectangular and unlined circular ducts are based on experimental data presented by Chaddock, Sabine, Ver, and Woods Fans (10,14,39,45). The algorithms for lined and rectangular and circular ducts are also based on experimental data. The algorithm for lined rectangular ducts is based on data by Kuntz, Kuntz and Hoover, and Machen and Haines (22,23,24). Multivariable regression analyses were performed on these

1-1

data to obtain a simple set of equations that can be programmed on a computer. The algorithm for lined circular ducts is based on data obtained from Bodley (8). Similar multivariable regression analyses were performed on this data. The algorithms for unlined and lined square elbows are based on information in the ASHRAE Handbook (4,37,38,40). The algorithm for acoustically lined circular radiused elbows is based on data obtained by Bodley (8). As with lined ducts, a multivariable regression analysis was performed on these data to obtain a set of equations that is easily implemented in a computer program. The algorithm for duct silencers is based solely on product data. This should be kept in mind when using this algorithm, as it would be best to use product data specific to the duct silencers being installed in the ductwork. The algorithm for duct branch sound power division is based on ASHRAE-sponsored work by Ver (44,46). Very little experimental data relative to the low-frequency sound attenuation associated with duct end reflection losses are available in the open literature. The algorithm for duct end reflection loss is based on work completed by Sandbakken, Pande, and Crocker (32). This work resulted in the technical basis for the duct end reflection correction that is part of AMCA Standard 300-85, "Reverberation Room Method for Sound Testing of Fans" (1). The algorithm for terminal volume regulation units is based solely on the 1984 ASHRAE Handbook (38). Most of the available data will be proprietary manufacturers' data. This algorithm should be used only as a general guide. Whenever possible, it is advisable to use manufacturers' data for determining the sound attenuation associated with terminal volume regulation units.

Algorithms that fall under the general heading of "Duct Breakout and Breakin" include: breakout and breakin of rectangular ducts, breakout and breakin of circular ducts, breakout and breakin of flat-oval ducts, and breakout and breakin of externally lagged rectangular ducts. All the algorithms under this general heading are based on work by Cummings, which is also summarized in the ASHRAE handbook (11,12,13,36).

The algorithms that are discussed under the heading "Sound Transmission in Indoor and Outdoor Spaces" include: sound attenuation through ceiling systems; receiver room sound corrections; sound transmission through mechanical equipment room walls, floor and ceiling; and sound transmission in outdoor spaces. The algorithm for the sound attenuation through ceiling systems is based on product data and the ASHRAE handbook (35,36). The coefficients for acoustical leaks and the quality of construction are somewhat subjective and should be used with caution. Algorithms for the receiver room correction are based on work that has been reported by Beranek, Schultz, and Thompson (2,33,42,43). The algorithm for sound transmission through mechanical equipment room walls and through holes in mechanical equipment room walls is based on work completed by Reynolds and Bledsoe (30). As is the case for ceilings, the coefficients for acoustical leaks and the quality of construction are somewhat subjective and should be used with caution. The algorithm for sound attenuation in outdoor environments is presented to give the designer a general idea of outdoor sound attenuation. Approximations are made for barriers and reflecting surfaces. Caution should be exercised in using this information if outdoor noise levels are critical. A more in-depth analysis should be conducted if this is the case.

Computer programs have been developed for each of the algorithms previously mentioned. An outline, a listing, and sample output for each of the programs appear in the appendices of this publication. The programs are programmed in basic. They are examples of how each algorithm can be set up for operation on a computer.

Endless time can be spent in developing computer programs that will contain all of the capabilities and error checking options that their many users desire them to have. The programs listed in the appendices are designed to be reasonably user-friendly and to give accurate results. They are not designed to contain all of the capabilities and error checking options that an integrated, commercial computer program on HVAC acoustics would be expected to have. It was not the intent of ASHRAE Research Project 556-RP to do this. However, the programs are written, so that, subject to the copyright restrictions associated with this publication, a user can incorporate them into an integrated computer program in the area of HVAC acoustics that has all of the capabilities and error checking options he or she desires to add.

SOME BASICS

CHAPTER 2

2.1 SOUND LEVELS

The most common parameter used to give an indication of "loudness" is the sound pressure level, L_p. The sound pressure level, L_p (dB), is defined as

$$L_p = 10 \log_{10} \left[\frac{p_{rms}^2}{p_{ref}^2} \right] \tag{2.1}$$

where p_{rms} is the root-mean-square value (rms) of acoustic pressure (Pa). p_{ref} is the reference sound pressure and has a value of 2×10^{-5} Pa or 0.0002 μbar. This amplitude was selected because it is the amplitude of the sound pressure that roughly corresponds to the threshold of hearing at a frequency of 1000 Hz. The intensity level, L_I (dB), is defined as

$$L_I = 10 \log_{10} \left[\frac{I}{I_{ref}} \right] \tag{2.2}$$

where I is acoustic intensity (W/m^2). I_{ref} is the reference intensity level and has a value of 10^{-12} W/m^2. The sound power level, L_W (dB), is defined as

$$L_W = 10 \log_{10} \left[\frac{W}{W_{ref}} \right] \tag{2.3}$$

where W is sound power (W). W_{ref} is the reference sound power and has a value of 10^{-12} watts.

With respect to HVAC systems, noise reduction, NR (dB), is

$$NR = L_{p(1)} - L_{p(2)} \tag{2.4}$$

where $L_{p(1)}$ is the sound pressure level (dB) of the sound entering a duct element and $L_{p(2)}$ is the sound pressure level (dB) of the sound coming out of the element. Insertion loss, IL (dB), is

$$IL = L_{p(w/o)} - L_{p(w)} \tag{2.5}$$

where $L_{p(w/o)}$ is the sound pressure level (dB) at a point without a specific duct element inserted and $L_{p(w)}$ is the sound pressure level (dB) at the same point with the duct element inserted. Transmission loss, TL (dB), is

$$TL = -10 \log_{10} \left[\frac{W_{out}}{W_{in}} \right] \tag{2.6}$$

where W_{in} is the sound power (W) of sound entering a duct element and W_{out} is the sound power (W) of the sound exiting the duct element.

Often it is necessary to add the sound pressure levels at a point in a room from several sound sources in the room, or it is necessary to add the sound power levels at a specific point in a duct system associated with different duct elements. When adding sound pressure or sound power levels, the total level, L_T (dB), is

$$L_T = 10 \log_{10} [SUM] \tag{2.7}$$

where

$$SUM = 10^{[L_{p1}/10]} + 10^{[L_{p2}/10]} + 10^{[L_{p3}/10]} + \ldots \tag{2.8}$$

when adding sound pressure levels, and

$$\text{SUM} = 10^{[L_{W1}/10]} + 10^{[L_{W2}/10]} + 10^{[L_{W3}/10]} + \ldots \tag{2.9}$$

when adding sound power levels. When examining the sound propagation in an HVAC system, it is necessary to subtract noise reduction, insertion loss, or transmission loss values from given values of sound power levels at different points in the system. When this is done,

$$L_{W2} = L_{W1} - NR \quad \text{or} \quad L_{W2} = L_{W1} - IL \quad \text{or} \quad L_{W2} = L_{W1} - TL \tag{2.10}$$

where L_{W1} is the sound power level before a duct element and L_{W2} is the sound power level after the element.

EXAMPLE 2.1

The following sound power levels exist at a point in a duct system, and the IL values are associated with a duct element that exists at the point.

	\multicolumn{8}{c}{1/1 Octave Band Center Frequency — Hz}
	63	125	250	500	1000	2000	4000	8000
L_{W1}	84	87	90	89	91	88	87	84
L_{W2}	74	95	93	98	99	92	88	87
L_{W3}	70	78	86	86	73	67	59	57
IL	6	11	17	22	28	32	24	20

Determine the sound power levels that exist after the duct element.

SOLUTION

The results are summarized below.

	63	125	250	500	1000	2000	4000	8000
$(W_1/W_{ref}) \times 10^{-8}$	2.51	5.01	10.00	7.94	12.59	6.31	5.012	2.512
$(W_2/W_{ref}) \times 10^{-8}$	0.25	31.62	19.95	63.10	79.43	15.85	6.310	5.012
$(W_3/W_{ref}) \times 10^{-8}$	0.10	0.63	3.98	3.98	0.20	0.05	0.008	0.005
SUM $\times 10^{-8}$	2.86	37.26	33.93	75.02	92.22	22.21	11.33	7.529
L_{WT}, dB	84.6	95.7	95.3	98.8	99.6	93.5	90.5	88.8
IL, dB	−6.0	−11.0	−17.0	−22.0	−28.0	−32.0	−24.0	−20.0
L_W, dB	78.6	84.7	78.3	76.8	71.6	61.5	66.5	68.8

EXAMPLE 2.2

The following sound pressure levels are given at a specified point within a room.

	\multicolumn{8}{c}{1/1 Octave Band Center Frequency — Hz}
	63	125	250	500	1000	2000	4000	8000
L_{p1}	74	78	77	77	75	77	75	66
L_{p2}	73	83	86	80	76	78	79	71
L_{p3}	84	89	88	86	89	88	85	78

Determine the total sound pressure level.

SOME BASICS

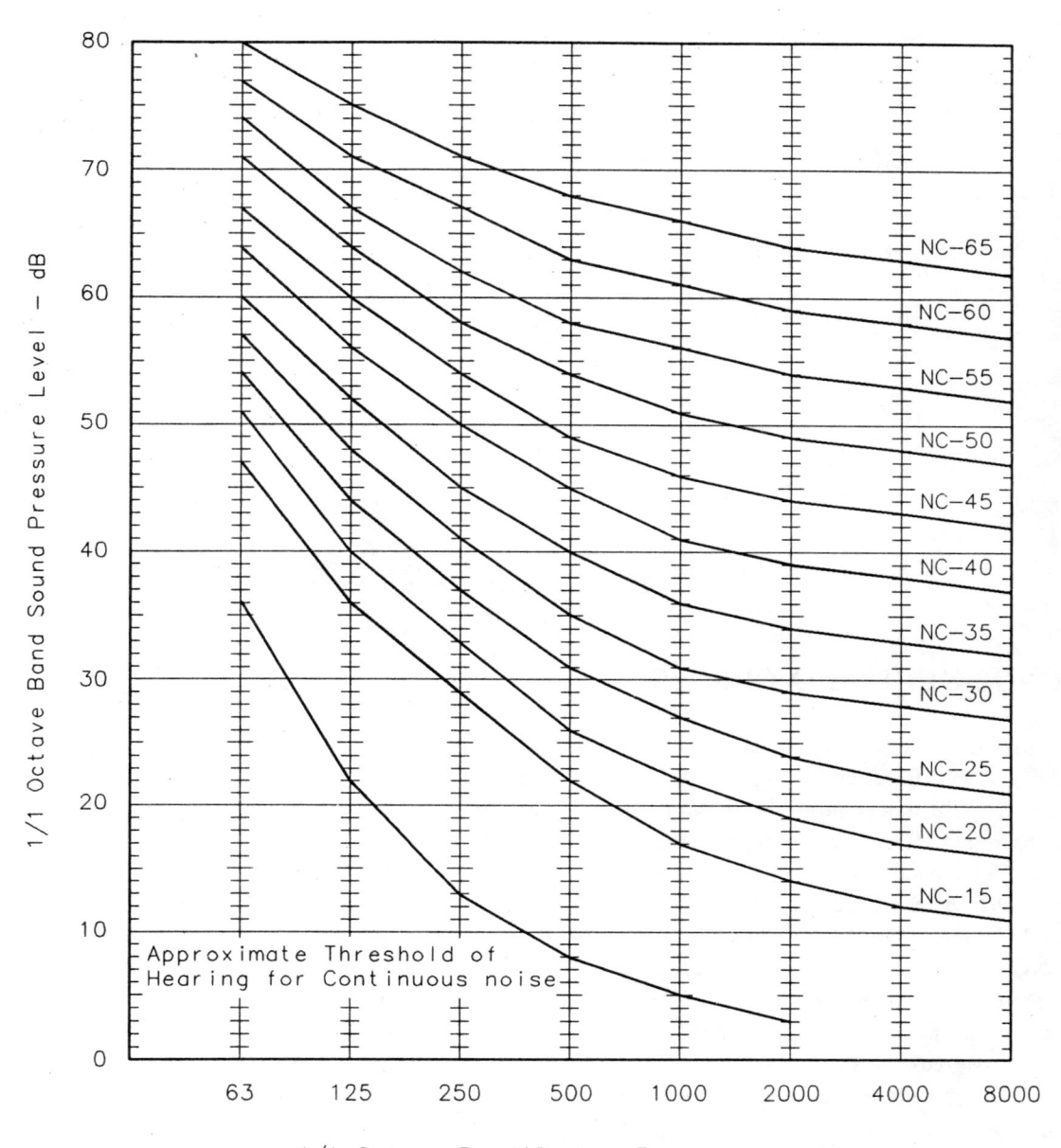

Figure 2.1 Noise Criterion Curves

SOLUTION

The results are summarized below.

	\multicolumn{8}{c}{1/1 Octave Band Center Frequency — Hz}
	63	125	250	500	1000	2000	4000	8000
$(p_{rms1}/p_{ref}) \times 10^{-7}$	2.51	6.31	5.01	5.01	3.16	5.01	3.16	0.40
$(p_{rms2}/p_{ref}) \times 10^{-7}$	2.00	19.95	39.81	10.00	4.00	6.31	7.94	1.26
$(p_{rms3}/p_{ref}) \times 10^{-7}$	25.12	79.43	63.10	39.81	79.43	63.10	31.62	6.31
SUM $\times 10^{-7}$	29.63	105.7	107.9	54.82	86.59	74.42	42.72	7.97
L_{pT}, dB	84.7	90.2	90.3	87.4	89.4	88.7	86.3	79.0

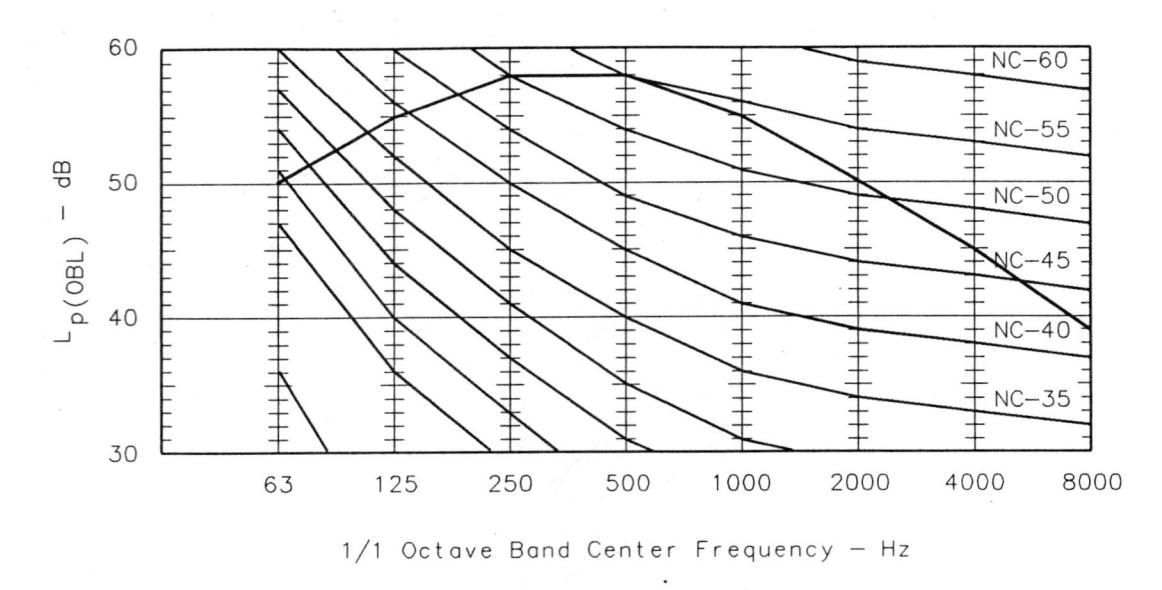

Figure 2.2 NC Level for example 2.3

2.2 NOISE CRITERIA CURVES

In an effort to simplify the task of an engineer in specifying acceptable acoustical environments for different activity areas, Beranek carried out a study that resulted in a set of noise criterion curves (3). These curves are a result of a statistical analysis of data from the subjective responses of many office workers to various acoustical environments. The noise ratings made by these workers were plotted against speech interference levels and loudness levels. The resulting speech interference and loudness level criteria were then translated into the noise criteria curves shown in Figure 2.1. These curves apply to steady noise and specify the maximum noise level permitted in each 1/1 octave band for a specified NC curve. For example, if the noise requirements for an activity area call for an NC 20 rating, the sound pressure levels in all eight 1/1 octave frequency bands must be less than the corresponding values for the NC 20 curve. Conversely, the NC rating of a given noise equals the highest penetration of any of the 1/1 octave band sound pressure levels into the curves. If the farthest penetration falls between two curves, the NC rating is the interpolated value between the two curves.

Table 2.1 shows the recommended NC levels for several activity areas (37). The lower NC levels in the table should be used in buildings where high-quality acoustical environments are desired. The upper levels should be used only for situations where economics or other conditions make use of the lower values impractical. Table 2.2 shows telephone use and listening conditions as a function of NC levels.

EXAMPLE 2.3

The following 1/1 octave band sound pressure levels were measured in a laboratory work area. What is the NC rating of the noise in the work area?

	63	125	250	500	1000	2000	4000	8000
L_p, dB	50	55	58	58	55	50	45	39

Figure 2.2 shows a plot of the above data relative to the NC curves. Since the 1/1 octave band sound pressure level in the 500 Hz 1/1 octave band penetrates to the NC 55 curve, the NC rating of the work area is NC 55.

Table 2.1 Recommended NC and RC(N) Levels for Different Indoor Activity Areas

Type of Area	NC or RC Level	Approx dBA	Type of Area	NC or RC Level	Approx dBA
RESIDENCES			**CHURCHES AND SCHOOLS**		
Private home (rural and suburban)	20–30	25–35	Sanctuaries	20–30	25–35
Private home (urban)	25–35	30–40	Libraries	30–40	35–45
Apartment house	30–40	35–45	Schools & classrooms	30–40	35–45
HOTELS			Laboratories	35–45	40–50
Individual rooms	30–40	35–45	Recreation halls	35–50	40–55
Ballroom, banquet rm	30–40	35–45	Corridors & halls	35–50	40–55
Halls, corridors, lobbies	35–45	40–50	**PUBLIC BUILDINGS**		
Garages	40–50	45–55	Libraries, museums	30–40	35–45
Kitchens, laundries	40–50	45–55	Court rooms	30–40	35–45
HOSPITALS & CLINICS			Post offices, lobbies	35–45	40–50
Private rooms	25–35	30–40	Gen. banking areas	35–45	40–50
Operating rooms	30–40	35–45	Washrooms, toilets	40–50	45–55
Wards, corridors	30–40	35–45	**RESTAURANTS, LOUNGES, CAFETERIAS**		
Laboratories	30–40	35–45	Restaurants	35–45	40–50
Lobbies, waiting rms	35–45	40–50	Cocktail lounges	35–40	40–45
Washrooms, toilets	40–50	45–55	Nightclubs	35–45	40–50
OFFICES			Cafeterias	40–50	45–55
Board rooms	20–30	25–35	**RETAIL STORES**		
Conference rooms	25–35	30–40	Clothing stores	35–45	40–50
Executive offices	30–40	35–45	Department stores (upper floors)	35–45	40–50
General offices	30–45	35–50	Department stores (main floors)	40–50	45–55
Reception rooms	30–45	35–50	Small retail stores	40–50	45–55
General open offices	35–45	40–50	Supermarkets	40–50	45–55
Drafting rooms	35–45	40–50	**INDOOR SPORTS ACTIVITIES**		
Halls & corridors	40–55	45–60	Coliseums	30–40	35–45
Tabulation and computation areas	40–50	45–55	Bowling alleys	35–45	40–50
AUDITORIUMS AND MUSIC HALLS			Gymnasiums	35–45	40–50
Concert, opera halls	15–25	20–30	Swimming pools	40–55	45–60
Sound record studios	15–25	20–30	**TRANSPORTATION (RAIL, BUSES, PLANES)**		
Legitimate theaters	25–30	30–40	Ticket sales offices	30–40	35–45
Multi-purpose halls	25–30	30–35	Lounges, waiting rms	35–50	40–55
Movie theaters	30–35	35–40			
TV audience studios	30–35	35–40			
Amphitheaters	30–35	35–40			
Lecture halls	30–35	35–40			
Planetariums	30–35	35–40			
Lobbies	35–45	40–50			

Table 2.2 Listening Conditions and Telephone Use as a Function of NC and RC(N) Levels

NC or RC Level	Environment	Telephone Use	Listening Conditions
<20	Silent	Excellent	Critical
20–30	Quiet	Excellent	Excellent
30–40	Moderately Noisy	Good	Good to Satisfactory
40–50	Noisy	Satisfactory	Satisfactory to Slightly Difficult
50–60	Very Noisy	Difficult	Difficult

2.3 ROOM CRITERION CURVES

HVAC noise is often the primary type of background noise that exists in many indoor areas. Experience has indicated that when HVAC background noise is present the use of NC levels has often resulted in a poor correlation between the calculated NC levels and an individual's subjective response to the corresponding background noise. The method used to determine NC levels is not sensitive to the way in which people subjectively respond to background noise. People can respond quite differently to background noise sources that have the same NC levels, but have different spectrum shapes. Noise that has a spectrum shape similar to the NC curves in Figure 2.1 is not very pleasant sounding. It will usually have both a low-frequency rumble and a high-frequency hiss. As a means of overcoming the above shortcomings associated with NC levels, Blazier introduced the room criterion (RC) curves shown in Figure 2.3 (7). According to Blazier, there are four factors that should be considered when assessing HVAC system background noise: (1) level, (2) spectrum shape or balance, (3) tonal content, and (4) temporal fluctuations.

There are two parts to determining the RC noise rating associated with HVAC background noise. The first is the calculation of a number that corresponds to the speech communication or masking properties of the noise. The second is designating the quality or character of the background noise. The procedure for determining the RC rating is:

1. Calculate the arithmetic average of the 1/1 octave band sound pressure levels in the 500 Hz, 1,000 Hz and 2,000 Hz 1/1 octave frequency bands. Round off to the nearest integer. This is the RC level associated with the background noise.

2. Draw a line that has a −5 dB/octave slope that passes through the calculated RC level at 1,000 Hz. For example, if the RC level is RC 32, the line will pass through a value of 32 at the 1,000 Hz 1/1 octave band. This value may not be equal to the value of the 1/1 octave band sound pressure level of the background noise in the 1,000 Hz 1/1 octave band.

3. Determine the subjective quality or character of the background noise.

According to Blazier, the subjective rating of background noise of similar RC levels can be classified as follows:

1. Neutral: Noise that is classified as neutral has no particular identity with frequency. It is usually bland and unobtrusive. Background noise that is neutral usually has a 1/1 octave band spectrum shape similar to the RC curves in Figure 2.3. If the 1/1 octave band data do not exceed the RC curve by 5 dB, the background noise is neutral and an "(N)" can be placed after the RC level.

2. Rumble: Noise that has a rumble has an excess of low-frequency sound energy. If any of the 1/1 octave band sound pressure levels below the 500 Hz 1/1 octave band are more than 5 dB above the RC curve associated with the background noise in a room, the noise will be judged to have a "rumbly" quality or character. If the background has a rumbly quality, place an "(R)" after the RC level.

3. Hiss: Noise that has an excess of high-frequency sound energy will have a "hissy" quality. If any of the 1/1 octave band sound pressure levels above the 500 Hz 1/1 octave band are more than 3 dB above the RC curve, the noise will be judged to have a hissy quality. If the background noise has a hissy quality, place an "(H)" after the RC level.

4. Tonal: Noise that has a tonal character usually contains a humming, buzzing, whining, or whistling sound. When a background sound has a tonal quality, it will generally have one 1/1 octave band in which the sound pressure level is noticeably higher than the other 1/1 octave bands. If the background noise has a tonal character, place a "T" after the RC level.

Background noise that has a 1/1 octave band spectrum that falls within the limiting boundaries identified with rumble and hiss and that has no tonal components is classified as neutral.

It is desirable for background noise to have a 1/1 octave band spectrum that has a neutral character or quality. If the noise spectrum is such that is has a rumbly, hissy, or tonal character, it will generally be judged to be objectionable. If the background noise has a neutral quality, the NC levels specified in Tables 2.1 and 2.2 can be used to indicate the desired RC(N) levels in different indoor activity areas.

EXAMPLE 2.4

The 1/1 octave band sound pressure levels of background noise in an office area are given below:

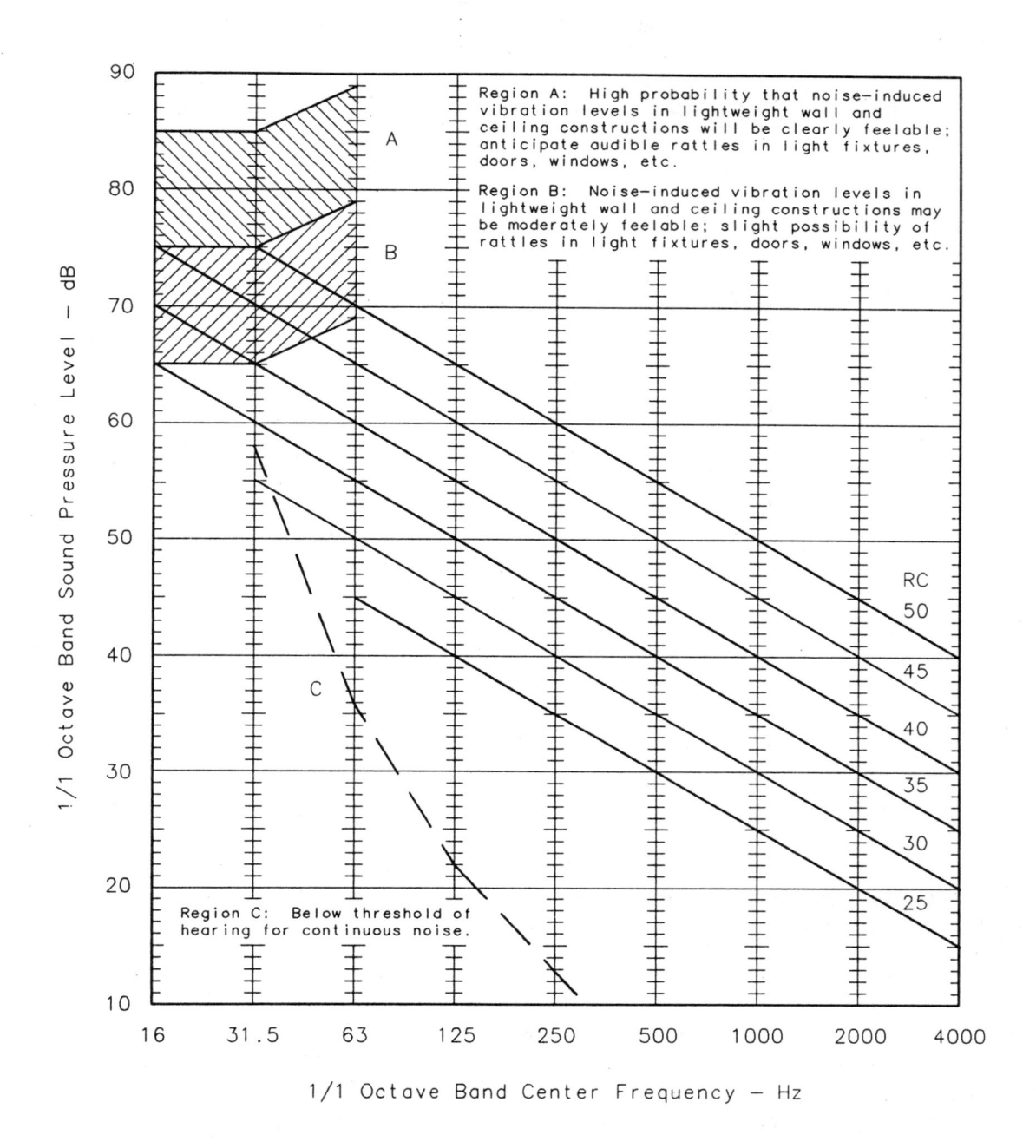

Figure 2.3 Room Criterion Curves

	1/1 Octave Band Center Frequency – Hz							
	31.5	63	125	250	500	1000	2000	4000
L_p, dB	70	62	54	46	40	33	27	20

Determine the RC level and the corresponding character of the noise.

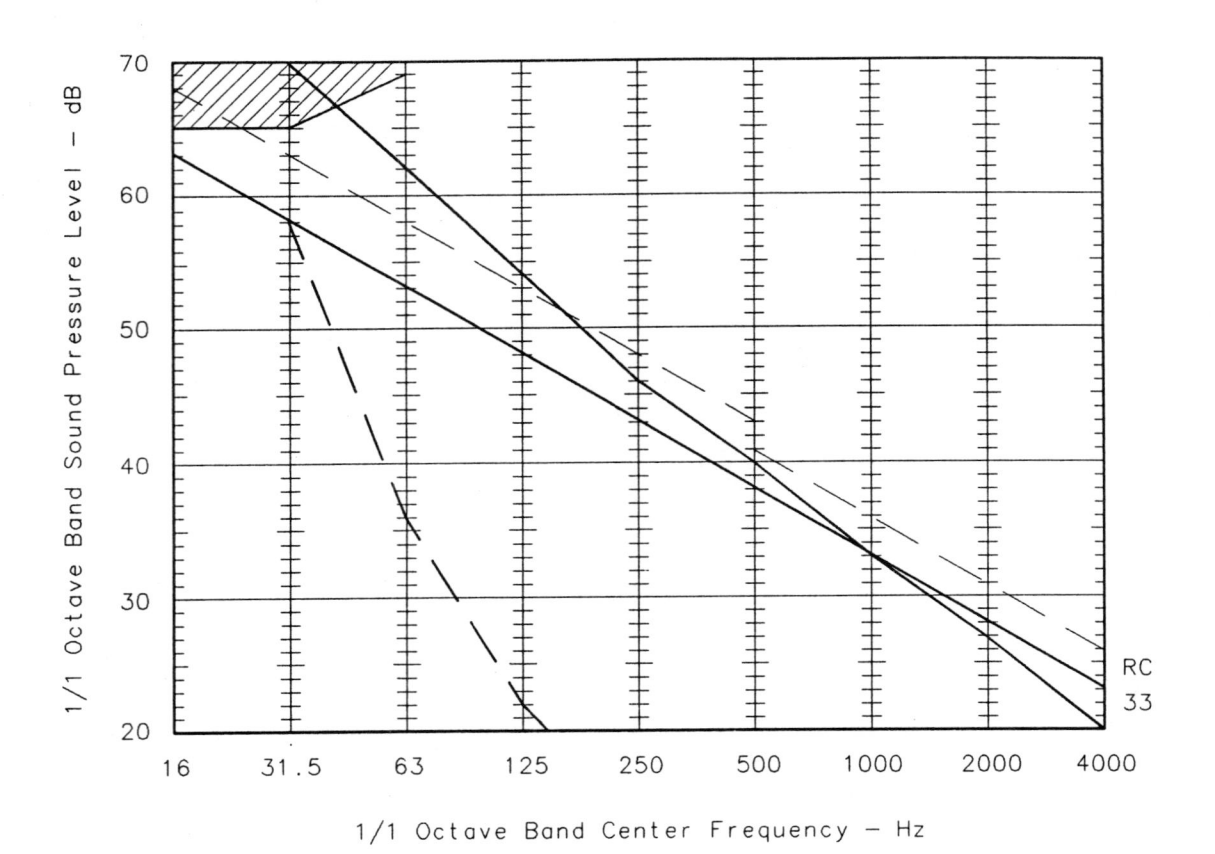

Figure 2.4 RC Level for example 2.4

SOLUTION

The RC level is determined by obtaining the arithmetic average of the 1/1 octave band sound pressure levels in the 500 Hz, 1,000 Hz, and 2,000 Hz 1/1 octave bands, or

$$RC = \frac{40 + 33 + 27}{3} \quad \text{or} \quad RC = 33.$$

Thus, the RC level is RC 33. The 1/1 octave band sound pressure levels for the background noise are plotted in Figure 2.4. The RC 33 curve (level in 1,000 Hz 1/1 octave band is 33 dB) is shown in the figure. A dashed line 5 dB above the RC 33 curve for frequencies below 500 Hz and a dashed line 3 dB above the RC 33 curve for frequencies above 500 Hz are also shown in the figure. An examination of the figure indicates that at frequencies below the 250 Hz 1/1 octave band, the 1/1 octave band sound pressure levels of the background noise are 5 dB or more above the RC 33 curve. Thus, the background noise has a rumbly character. The 1/1 octave band sound pressure levels above 500 Hz are equal to or below the RC 33 curve, so there is no problem at these frequencies. The RC rating of the background noise is RC 33(R).

EQUIPMENT SOUND LEVELS
CHAPTER 3

3.1 FANS

The sound power generation of a given fan performing a specific task is best obtained from the fan manufacturer's test data. Manufacturers' test data should be obtained from either AMCA Standard 300–85, "Reverberent Room Method for Sound Testing of Fans," or ANSI/ASHRAE Standard 68–1986/ANSI/AMCA Standard 330–86, "Laboratory Method of Testing In–Duct Sound Power Measurement Procedure for Fans" (52). However, when such data are not available, the 1/1 octave band sound power levels for various fans can be estimated by the procedure described below.

Fan noise can be rated in terms of the specific sound power levels. This is defined as the sound power level generated by a fan operating at a volume flow rate of 1 ft^3/min (cfm) and a static pressure of 1 in. of H_2O. By reducing all fan noise data to this common denominator, the specific sound power level serves as a basis for direct comparison of the 1/1 octave band sound power levels of different types of fans and as a basis for a conventional method of estimating the fan sound power levels of fans under actual operating conditions. With respect to specific sound power levels, small fans tend to have higher values than large fans. While the size divisions shown in Table 3.1 are somewhat arbitrary, these divisions are practical for estimating fan noise.

Fans generate a tone at the blade passage frequency. To account for this, the sound power level in the 1/1 octave band in which the blade passage frequency occurs is increased by a specified amount. The number of decibels to be added to this 1/1 octave band is called the blade frequency increment (BFI). Table 3.2 gives an estimate of the 1/1 octave band for different types of fans in which the blade passage frequency occurs and the corresponding blade passage frequency increment. For a more accurate estimate of the blade passage frequency, B_f, the following equation can be used:

$$B_f = \frac{rpm}{60} \times \text{no. of blades} \tag{3.1}$$

where rpm is the rotational speed of the fan in revolutions per minute.

The specific sound power levels for fan total sound power given in Table 3.1 are for fans operating at a point of operation where the volume flow rate equals 1 cfm (0.472 L/s) and the static pressure is 1 in. H_2O (249 Pa). Equation (3.2) is used to calculate the inlet or outlet fan sound power levels corresponding to a specific point of operation.

$$L_W = K_W + 10 \log_{10}\left[\frac{Q}{Q_1}\right] + 20 \log_{10}\left[\frac{P}{P_1}\right] + C \tag{3.2}$$

where L_W is the estimated sound power level of the fan in dB, K_W is the specific sound power level in dB from Table 3.1, Q is the flow rate in cfm (L/s), Q_1 is 1 cfm (0.472 L/s), P is the pressure drop in in. H_2O (Pa), P_1 is 1 in. H_2O (249 Pa), and C is the correction factor in dB for the case where the point of fan operation is other than the point of peak efficiency. Values for C are obtained from Table 3.3.

EXAMPLE 3.1

A forward–curved fan supplies 10,000 cfm of air at a static pressure of 1.5 in. H_2O. It has 24 blades and operates at 1,175 rpm. The fan has a peak efficiency of 85%. The fan brake horsepower is 3 BHp. Determine the fan total sound power levels.

SOLUTION

$$L_W = K_W + 10 \log_{10}[Q] + 20 \log_{10}[P] + C$$

$$\text{Operating efficiency} = E1 = \frac{\text{volume flow} \times \text{static pressure}}{6356 \times BHp} \times 100$$

EQUIPMENT SOUND POWER

Table 3.1 Specific Sound Power Levels, K_W, for Fan Total Sound Power (36,54)

Fan Type		63	125	250	500	1000	2000	4000	8000
Centrifugal:									
Airfoil, Backward Curved, Backward Inclined	Wheel Diameter (in)								
	≥ 36	40	40	39	34	30	23	19	17
	< 36	45	45	43	39	34	28	24	19
Forward Curved	All	53	53	43	36	36	31	26	21
Radial	Total Press (in H_2O)								
Material Wheel	4–10	56	47	43	39	37	32	29	26
Medium Press	6–15	58	54	45	42	38	33	29	26
High Press	15–60	61	58	53	48	46	44	41	38
Vaneaxial:	Hub Ratio								
	0.3–0.4	49	43	43	48	47	45	38	34
	0.4–0.6	49	43	46	43	41	36	30	28
	0.6–0.8	53	52	51	51	49	47	43	40
Tubeaxial:	Wheel Diameter (in)								
	≥ 40	51	46	47	49	47	46	39	37
	< 40	48	47	49	53	52	51	43	40
Propeller:									
General ventilation and Cooling towers	All	48	51	58	56	55	52	46	42

Table 3.2 Blade Frequency Increments (BFI) (36)

Fan Type	1/1 Octave Band in which BFI occurs	BFI dB
Centrifugal		
Airfoil, backward curved backward inclined	250 Hz	3
Forward curved	500 Hz	2
Radial blade, pressure blower	125 Hz	8
Vaneaxial	125 Hz	6
Tubeaxial	63 Hz	7
Propeller		
General Ventilation and Cooling Tower	63 Hz	5

Table 3.3 Correction Factor, C, for Off–Peak Operation (36)

Static Efficiency % of Peak	Correction Factor dB
90 to 100	0
85 to 89	3
75 to 85	6
65 to 74	9
55 to 64	12
50 to 54	15
below 50	16

$$E1 = \frac{10{,}000 \times 1.5}{6{,}356 \times 3} \times 100 = 79\,\% \qquad \text{Peak efficiency} = E2 = 85\,\%$$

$$\%\text{ of static efficiency} = \frac{E1}{E2} \times 100 = \frac{79}{85} \times 100 = 93\%$$

From Table 3.3, the correction for off–peak operation = 0 dB.

$$L_W = K_W + 10\log_{10}[10{,}000] + 20\log_{10}[1.5] + 0.0 = K_W + 44.$$

$$B_f = \frac{1175}{60} \times 24 = 470 \text{ Hz, which is in the 500 Hz 1/1 octave band.}$$

From Table 3.2, blade frequency increment = 2 dB. The results are tabulated below.

	\multicolumn{8}{c}{1/1 Octave Band Center Frequency – Hz}
	63	125	250	500	1000	2000	4000	8000
K_W – Table 3.1, dB	53	53	43	36	36	31	26	21
Equation (3.2), dB	44	44	44	44	44	44	44	44
Table 3.2, dB				2				
$L_{w(OBL)}$ (dB)	97	97	87	82	80	75	70	65

3.2 CHILLERS

Little data are available that can be used to accurately estimate the sound levels associated with chillers. This is partly associated with the size of chillers. Chillers are usually large in size; and as a result, it is often necessary to measure their sound levels in the equipment rooms in which they are installed. The acoustic properties of these rooms are sometimes difficult to accurately characterize; and there are often other noise sources present in the rooms in which the chiller sound levels are measured.

Chillers have several components that generate and radiate sound. However, the primary sound sources are usually the compressor or compressors associated with the chiller and the drive units used to run the compressors. The drive units are usually electric motors and sometimes steam turbines. It is not uncommon for large chillers to be made up of assemblies that have two or more smaller compressors. The sound levels given in this section for chillers are specified in terms of sound pressure levels, normalized to a distance of 3 ft from the acoustic center of the machine. Often the acoustic center of a machine is not clearly defined. When this is the case, the acoustic center is usually assumed to be at the geometric center of the machine.

Whenever possible, it is advisable to use manufacturers' published data. Manufacturers' data should be obtained in accordance with ARI Standard 575–87, "Standard for Method of Measuring Machinery Sound within Equipment Rooms" (50). A rough approximation of chiller sound pressure levels at a distance of 3 ft from the acoustic center of a chiller can be determined by using Equation (3.3) or (3.4) in conjunction with Table 3.4 (36). Equations (3.3) and (3.4) give the A weighted sound pressure level of a chiller as a function of tons refrigeration, TR.

Chillers with centrifugal compressors:

$$L_{p(A)}\,(@\,3\text{ ft}) = 60 + 11\log_{10}(TR) \quad \text{dBA}. \tag{3.3}$$

Chillers with reciprocating compressors:

$$L_{p(A)}\,(@\,3\text{ ft}) = 71 + 9\log_{10}(TR) \quad \text{dBA}. \tag{3.4}$$

Table 3.4 gives the correction values to be subtracted from the A weighted sound pressure level to obtain the corresponding 1/1 octave band sound pressure levels.

Table 3.4 Correction Values for Obtaining Chiller 1/1 Octave Band Sound Pressure Levels (36)

Values to be Subtracted from A Weighted L_p — dB	63	125	250	500	1000	2000	4000
Chiller with:							
Centrifugal Compressor Internal Geared Medium to Full Load	8	5	6	7	8	5	8
Centrifugal Compressor Direct Drive Medium to Full Load	8	6	7	3	4	7	12
Centrifugal Compressor >1000 Ton Medium to Full Load	11	11	8	8	4	6	13
Chiller with:							
Reciprocating Compressor All Loads	19	11	7	1	4	9	14

(Column headers: 1/1 Octave Band Center Frequency — Hz)

EXAMPLE 3.2

Determine the 1/1 octave band sound pressure levels of a 100-ton reciprocating chiller.

SOLUTION

Since the chiller is a reciprocating chiller, use Equation (3.4) to determine the A weighted sound pressure level. Thus,

$$L_{p(A)} (@ 3 \text{ ft}) = 71 + 9 \log_{10} [100] = 89 \text{ dBA}.$$

The results are tabulated below.

	63	125	250	500	1000	2000	4000
L_p (@ 3 ft), dBA	89	89	89	89	89	89	89
Table 3.4, dB	−19	−11	−7	−1	−4	−9	−14
L_p (@ 3 ft), dB	70	78	82	88	85	80	75

(Column headers: 1/1 Octave Band Center Frequency — Hz)

DUCT ELEMENT REGENERATED SOUND POWER — CHAPTER 4

4.1 DAMPERS

The 1/1 octave band sound power level of the noise generated by single or multi–blade dampers can be predicted by Equation (4.1) (36,44).

$$L_W(f_o) = K_D + 10 \log_{10}\left[\frac{f_o}{63}\right] + 50 \log_{10}\left[U_c\right] + 10 \log_{10}[S] + 10 \log_{10}[DH] \qquad (4.1)$$

where f_o is the 1/1 octave band center frequency (Hz), U_c is the flow velocity (ft/sec) in the constricted part of the flow field determined according to Equation (4.4), S is the cross–sectional area (ft^2) of the duct, DH is the duct height (ft) normal to the damper axis, and K_D is the characteristic spectrum (Figure 4.1). Figure 4.2 shows a schematic of a single–blade damper. The regenerated sound power levels associated with dampers are obtained as follows:

Step 1: Determine the total pressure loss coefficient, C.

$$C = 15.9 \times 10^6 \frac{\Delta P}{(Q/S)^2} \qquad (4.2)$$

where Q is the volume flow rate (ft^3/min), ΔP is the total pressure loss (in. H$_2$O) across the damper, and S is the duct cross–sectional area (ft^2).

Step 2: Determine the blockage factor, BF.
For multi–blade dampers:

$$BF = \frac{(\sqrt{C} - 1)}{(C - 1)} \qquad \text{If C = 1 then BF = 0.5.} \qquad (4.3a)$$

For single–blade dampers:

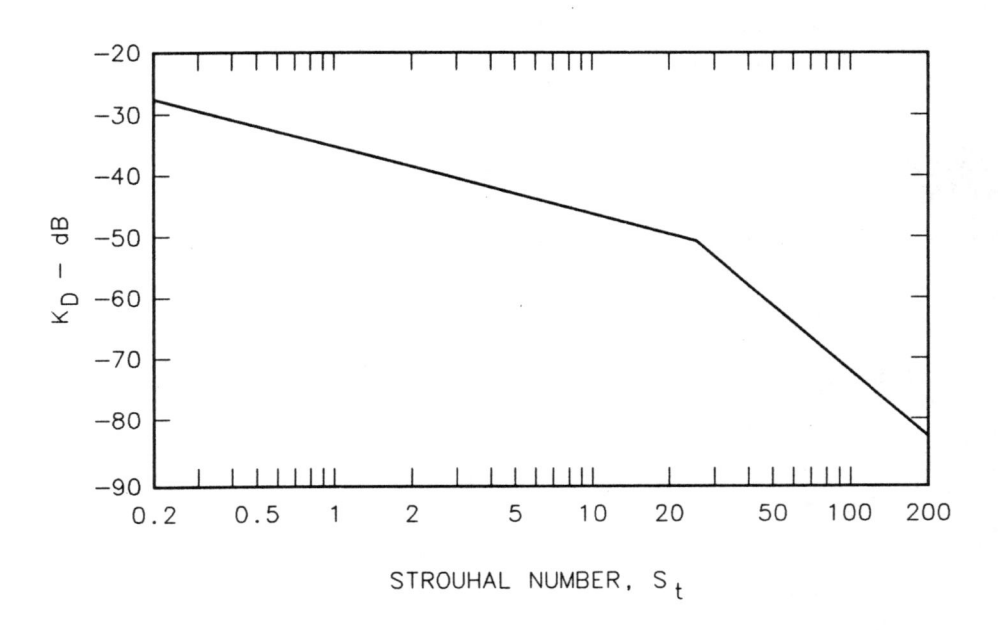

Figure 4.1 Characteristic Spectrum, K_D, for Dampers

$$BF = \begin{array}{l} \dfrac{(\sqrt{C} - 1)}{(C - 1)} \quad \text{for } C < 4 \\ 0.68 \cdot C^{-0.15} - 0.22 \quad \text{for } C > 4 \end{array} \qquad (4.3b)$$

Step 3: Determine the flow velocity, U_c, in the damper constriction.

$$U_c = 0.0167 \cdot \frac{Q}{S \cdot BF} \quad \text{(ft/s)}. \qquad (4.4)$$

Step 4: Determine the Strouhal number, S_t.
The Strouhal number that corresponds to the 1/1 octave band center frequencies is given by

$$S_t = \frac{f_o \, DH}{U_c}. \qquad (4.5)$$

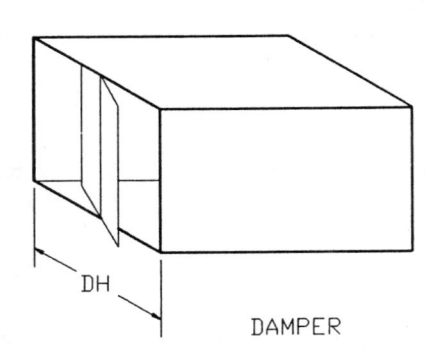

Figure 4.2 Damper

Step 5: Determine the characteristic spectrum, K_D.

The characteristic spectrum is the same for all dampers and duct sizes if plotted as a function of the Strouhal frequency. The characteristic spectrum, K_D, is obtained from Figure 4.1 or from

$$K_D = \begin{array}{l} -36.3 - 10.7 \, \text{Log}_{10}(S_t) \quad \text{for } S_t \leq 25 \\ -1.1 - 35.9 \, \text{Log}_{10}(S_t) \quad \text{for } S_t > 25 \end{array} \qquad (4.6)$$

All the required information is now available for calculating the 1/1 octave band sound power levels predicted by Equation (4.1).

EXAMPLE 4.1

Determine the 1/1 octave band sound power levels associated with a multi-blade damper positioned in a 12 in. by 12 in. duct. The pressure drop across the damper is 0.5 in. H_2O and the volume flow rate in the duct is 4,000 ft^3/min.

SOLUTION

From the given data:
$Q = 4,000 \text{ ft}^3/\text{min}$
$S = 1 \text{ ft} \times 1 \text{ ft} = 1 \text{ ft}^2$
$\Delta P = 0.5 \text{ in. } H_2O$
$DH = 1 \text{ ft}$

Step 1: Determine the total pressure loss coefficient, C.

$$C = 15.9 \times 10^6 \cdot \frac{0.5}{(4000/1)^2} = 0.5.$$

Step 2: Determine the blockage factor, BF.

$$BF = \frac{(\sqrt{0.5} - 1)}{(0.5 - 1)} = 0.585.$$

Step 3: Determine the constricted flow velocity, U_c.

$$U_c = 0.0167 \cdot \frac{4000}{1.0 \cdot 0.585} = 114 \text{ (ft/s)}.$$

The results are tabulated below.

	\multicolumn{8}{c}{1/1 Octave Band Center Frequency — Hz}
	63	125	250	500	1000	2000	4000	8000
S_t	0.55	1.1	2.2	4.4	8.8	17.6	35.1	70.2
K_D, dB	−33.5	−36.7	−40.0	−43.2	−46.4	−49.6	−56.6	−67.4
$10 \log_{10} [f_o/63]$, dB	0.0	3.0	6.0	9.0	12.0	15.0	18.0	21.0
$50 \log_{10} [U_c]$, dB	102.8	102.8	102.8	102.8	102.8	102.8	102.8	102.8
$10 \log_{10} [S]$, dB	0.0	0.0	0.0	0.0	0.0	0.0	0.0	0.0
$10 \log_{10} [D]$, dB	0.0	0.0	0.0	0.0	0.0	0.0	0.0	0.0
$L_W(f_o)$, dB	69.3	69.1	68.8	68.6	68.4	68.2	64.2	56.4

4.2 ELBOWS FITTED WITH TURNING VANES

The 1/1 octave band sound power levels associated with the noise generated by elbows fitted with turning vanes can be predicted if the total pressure drop across the blades is known or can be estimated (36,44). The method that is presented applies to any elbow that has an angle between 60° and 120°. The 1/1 octave band sound power levels generated by elbows with turning vanes are given by

$$L_W(f_o) = K_T + 10 \log_{10} \left[\frac{f_o}{63}\right] + 50 \log_{10} [U_c] + 10 \log_{10} [S] + 10 \log_{10} [CD] + 10 \log_{10} [n] \quad (4.7)$$

where f_o is the 1/1 octave band center frequency (Hz), U_c is the flow velocity (ft/sec) in the constricted part of the flow field between the blades determined from Equation (4.10), S is the cross-sectional area (ft^2) of the duct, CD is the cord length (in.) of a typical vane, n is the number of turning vanes, and K_T is the characteristic spectrum (Figure 4.3). In addition to the above parameters, it is also necessary to know the duct height, DH (ft), normal to the turning vane length. The regenerated sound power levels associated with elbows with turning vanes are obtained as follows:

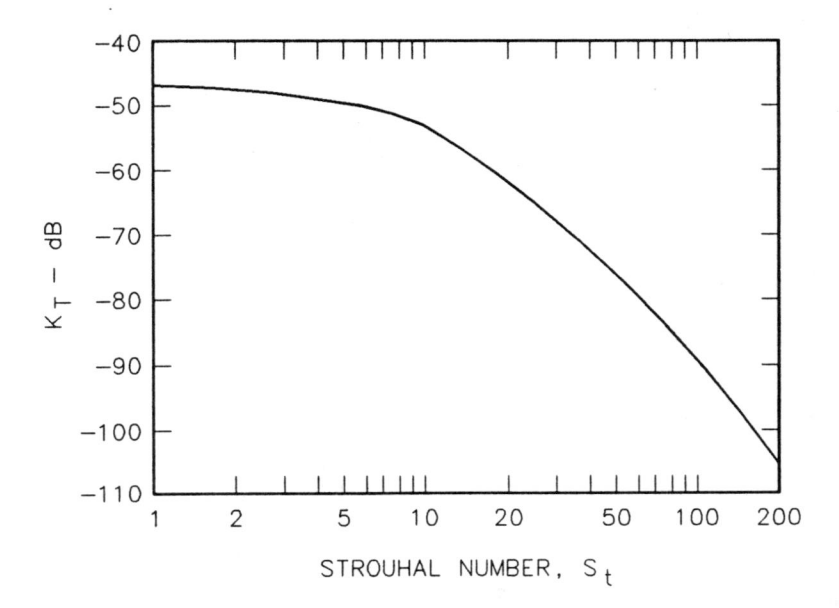

Figure 4.3 Characteristic Spectrum, K_T, for Elbows Fitted with Turning Vanes

DUCT ELEMENT REGENERATED SOUND POWER

Figure 4.4 90° Elbow with Turning Vanes

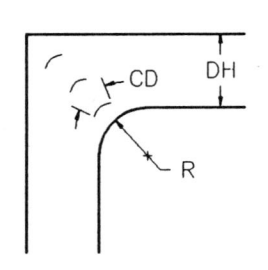

Step 1: Determine the total pressure loss coefficient, C.

$$C = 15.9 \times 10^6 \cdot \frac{\Delta P}{(Q/S)^2} . \qquad (4.8)$$

Step 2: Determine the blockage factor, BF.

$$BF = \frac{(\sqrt{C} - 1)}{(C - 1)} . \qquad (4.9)$$

Step 3: Determine the flow velocity, U_c (ft/sec), in the turning vane constriction.

$$U_c = 0.0167 \cdot \frac{Q}{S \cdot BF} . \qquad (4.10)$$

Step 4: Determine the Strouhal number, S_t.

$$S_t = \frac{f_o \, DH}{U_c} . \qquad (4.11)$$

Step 5: Determine the characteristic spectrum, K_T.

The characteristic spectrum is the same for any elbow fitted with turning vanes if plotted as a function of the Strouhal number. The characteristic spectrum is obtained from Figure 4.3 or from

$$K_T = -47.5 - 7.69 \left[\log_{10} (S_t) \right]^{2.5} . \qquad (4.12)$$

All the required information is now available for calculating the 1/1 octave band sound power levels predicted by Equation (4.7).

EXAMPLE 4.2

A 90° elbow of a 20 in. by 20 in. duct is fitted with five turning vanes that have a cord length of 7.9 in. The volume flow rate is 8,500 ft³/min, and the corresponding pressure loss across the turning vanes is 0.16 in. H₂O. Determine the resulting 1/1 octave band sound power levels.

SOLUTION

From the given data:
Q = 8,500 ft³/min
S = 2.78 ft²
CD = 7.9 in.

ΔP = 0.16 in. H₂O
DH = 1.64 ft
n = 5

Step 1: Determine total pressure loss coefficient, C.

$$C = 15.9 \times 10^6 \cdot \frac{0.16}{(8500/2.78)^2} = 0.27 .$$

Step 2: Determine the blockage factor, BF.

$$BF = \frac{(\sqrt{0.27} - 1)}{(0.27 - 1)} = 0.66 .$$

Step 3: Determine the constricted flow velocity, U_c.

$$U_c = 0.0167 \cdot \frac{8500}{2.78 \cdot 0.66} = 77.4 \text{ (ft/s)} .$$

The results are tabulated below.

	\multicolumn{8}{c}{1/1 Octave Band Center Frequency — Hz}
	63	125	250	500	1000	2000	4000	8000
S_t	1.3	2.6	5.3	10.6	21.2	42.4	84.8	167.5
K_T, dB	−47.5	−48.4	−50.9	−55.7	−63.1	−73.5	−87.0	−104.2
$10 \log_{10} [f_o/63]$, dB	0.0	3.0	6.0	9.0	12.0	15.0	18.0	21.0
$50 \log_{10} [U_c]$, dB	94.4	94.4	94.4	94.4	94.4	94.4	94.4	94.4
$10 \log_{10} [S]$, dB	4.4	4.4	4.4	4.4	4.4	4.4	4.4	4.4
$10 \log_{10} [CD]$, dB	9.0	9.0	9.0	9.0	9.0	9.0	9.0	9.0
$10 \log_{10} [n]$, dB	7.0	7.0	7.0	7.0	7.0	7.0	7.0	7.0
$L_W(f_o)$, dB	67.3	69.4	69.9	68.1	63.7	56.3	45.8	31.6

4.3 JUNCTIONS AND TURNS

Equation (4.13) has been developed as a means to predict the regenerated sound power levels in a branch duct associated with air flowing in duct turns and junctions (36,44). Equation (4.13) applies to 90° elbows without turning vanes, X–junctions, T–junctions, and 90° branch takeoffs (Figure 4.5).

$$L_W(f_o)_b = L_b(f_o) + \Delta r + \Delta T . \tag{4.13}$$

$L_b(f_o)$ is given by

$$L_b(f_o) = K_J + 10 \log_{10} \left[\frac{f_o}{63}\right] + 50 \log_{10} \left[U_B\right] + 10 \log_{10} \left[S_B\right] + 10 \log_{10} \left[D_B\right] \tag{4.14}$$

where f_o is the 1/1 octave band center frequency (Hz), D_B is the equivalent diameter (ft) of the branch duct, U_B is the flow velocity (ft/s) in the branch duct, S_B is the cross–sectional area (ft²) of the branch duct, and K_J is the characteristic spectrum (Figure 4.6). If the branch duct is circular, D_B is the duct diameter. If the branch duct is rectangular, D_B is obtained from

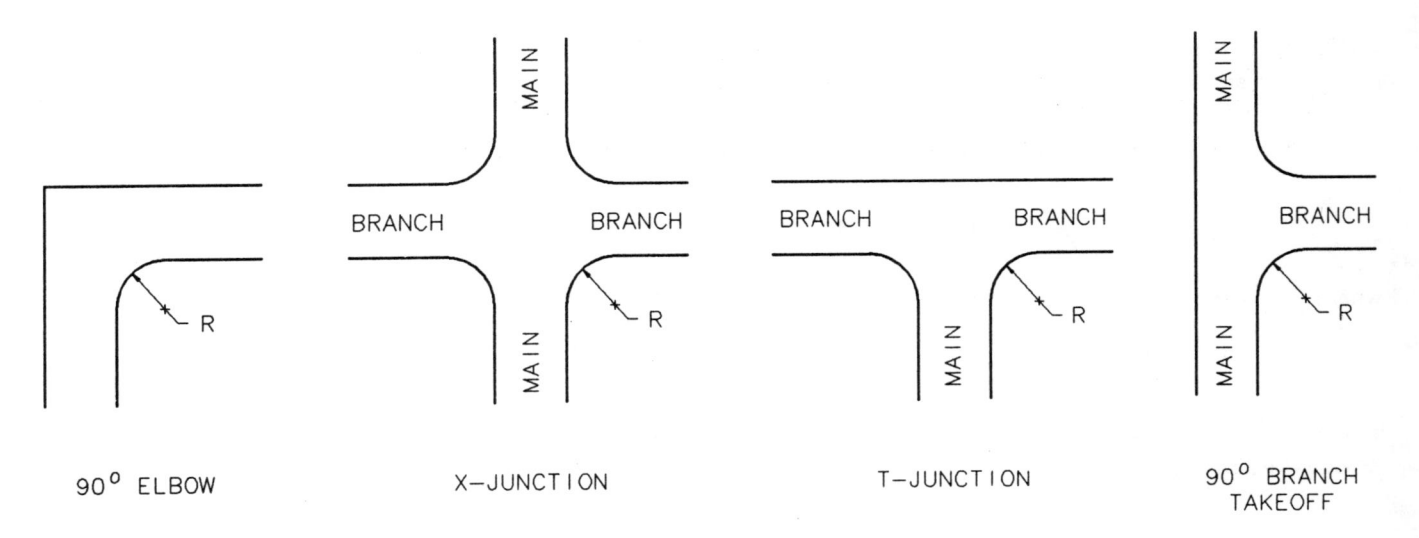

Figure 4.5 Elbows, Junctions, and Branch Takeoffs

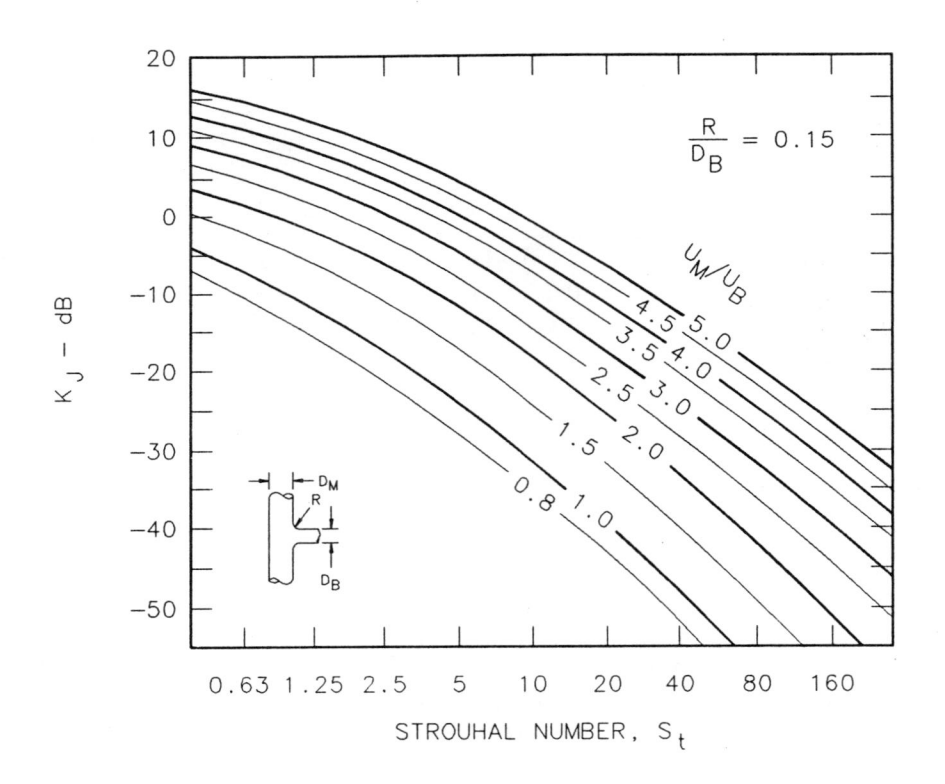

Figure 4.6 Characteristic Spectrum, K_J, for Junctions

$$D_B = \left[\frac{4 \cdot S_B}{\pi}\right]^{1/2}. \tag{4.15}$$

The corresponding flow velocity (ft/s), U_B, is given by

$$U_B = \frac{Q_B}{S_B \cdot 60} \tag{4.16}$$

where Q_B is the volume flow rate (ft³/min) in the branch. D_M (ft) and U_M (ft/s) for the main duct are obtained in manners similar to those implied by Equations (4.15) and (4.16).

Δr in Equation (4.13) is the correction term that quantifies the effect of the size of the radius of the bend or elbow associated with the turn or junction. Δr is obtained from Figure 4.7(a) or from

$$\Delta r = \left[1.0 - \frac{RD}{0.15}\right] \cdot \left[6.793 - 1.86 \log_{10}\left[S_t\right]\right] \tag{4.17}$$

where RD is the rounding parameter and S_t is the Strouhal number. RD is specified by

$$RD = \frac{R}{12 \cdot D_B} \tag{4.18}$$

where R is the radius (in.) of the bend or elbow associated with the turn or junction and D_B is defined above. The Strouhal number is given by

$$S_t = \frac{f_o D_B}{U_B}. \tag{4.19}$$

ΔT in Equation (4.13) is a correction factor for upstream turbulence. This correction is only applied when

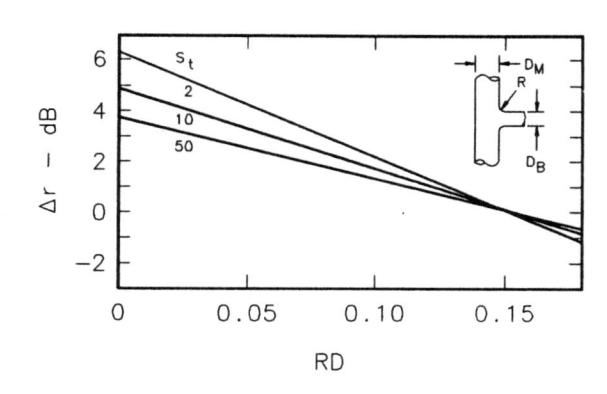

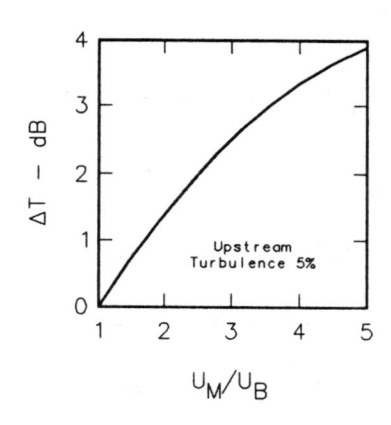

Figure 4.7 Correction Factors for Corner Rounding and for Upstream Turbulence

there are dampers, elbows, or branch takeoffs upstream within five main duct diameters of the turn or junction being examined. ΔT is obtained from Figure 4.7(b) or from

$$\Delta T = -1.667 + 1.8 \cdot m - 0.133 \cdot m^2 \tag{4.20}$$

where m is the velocity ratio that is specified by

$$m = \frac{U_M}{U_B}. \tag{4.21}$$

U_M is the flow velocity in the main duct before the turn or junction and U_B is the flow velocity in the branch duct after the turn or junction.

The characteristic spectrum, K_J, in Equation (4.14) is obtained from Figure 4.6 or from

$$K_J = -21.6 + 12.388 \cdot m^{0.673} - 16.482 \cdot m^{-0.303} \log_{10}\left[S_t\right] - 5.047 \cdot m^{-0.254} \left(\log_{10}\left[S_t\right]\right)^2. \tag{4.22}$$

The regenerated sound power levels in a branch duct that is associated with a turn or junction are obtained as follows:

Step 1: Obtain or determine the values of D_B and D_M.
Step 2: Determine the values of U_B and U_M.
Step 3: Determine the ratios D_M/D_B and m.
Step 4: Determine the rounding parameter, RD.
Step 5: Determine the Strouhal number, S_t.
Step 6: Determine the value of Δr.
Step 7: If turbulence is present, determine the value of ΔT.
Step 8: Determine the characteristic spectrum, K_J.
Step 9: Determine the value of the branch sound power levels, $L_w(f_o)_b$.
Step 10: Specify the type of turn or junction and determine the main duct sound power levels, $L_w(f_o)_m$, using Equations (4.23), (4.24), (4.25), or (4.26).

The related 1/1 octave band sound power levels of the noise generated in the main duct is given by the following equations:

X–Junction:

$$L_W(f_o)_m = L_W(f_o)_b + 20 \log_{10}\left[\frac{D_M}{D_B}\right] + 3 \qquad (4.23)$$

T–Junction:

$$L_W(f_o)_m = L_W(f_o)_b + 3 \qquad (4.24)$$

90° Elbow without Turning Vanes:

$$L_W(f_o)_m = L_W(f_o)_b \qquad (4.25)$$

90° Branch Takeoff:

$$L_W(f_o)_m = L_W(f_o)_b + 20 \log_{10}\left[\frac{D_M}{D_B}\right] \qquad (4.26)$$

EXAMPLE 4.3: X–Junction

Determine the regenerated sound power levels associated with an X–junction that exist in the branch and main ducts given the following information:

  Main Duct: Rectangular — 12 in. by 36 in., volume flow rate — 12,000 cfm
  Branch Duct: Rectangular — 10 in. by 10 in., volume flow rate — 1,200 cfm
  Radius of bend or elbow: 0.0 in.
  No dampers, elbows, or branch takeoffs are within five main duct diameters of the junction.

SOLUTION

Step 1: Determine the values of D_B and D_M.

$$D_M = \left[\frac{4 \cdot 12 \cdot 36}{\pi \cdot 144}\right]^{1/2} = 1.95 \text{ ft.} \qquad D_B = \left[\frac{4 \cdot 10 \cdot 10}{\pi \cdot 144}\right]^{1/2} = 0.94 \text{ ft.}$$

Step 2: Determine the values of U_B and U_M.

$$U_M = \frac{12000 \cdot 144}{12 \cdot 36 \cdot 60} = 66.67 \text{ ft/s}. \qquad U_B = \frac{1200 \cdot 144}{10 \cdot 10 \cdot 60} = 28.80 \text{ ft/s}.$$

Step 3: Determine the ratios, D_m/D_B and m.

$$\frac{D_M}{D_B} = \frac{1.95}{0.94} = 2.06. \qquad m = \frac{66.67}{28.80} = 2.31.$$

Step 4: Determine the rounding parameter, RD.

$$RD = \frac{0}{12 \cdot 0.94} = 0.$$

The results are tabulated below.

DUCT ELEMENT REGERERATED SOUND POWER

	63	125	250	500	1000	2000	4000	8000
		1/1 Octave Band Center Frequency — Hz						
S_t	2.0	4.1	8.2	16.3	32.6	65.3	130.6	261.2
K_J, dB	−4.2	−9.1	−14.9	−21.3	−28.5	−36.4	−45.1	−54.5
$10 \log_{10} [f_o/63]$, dB	0.0	3.0	6.0	9.0	12.0	15.0	18.0	21.0
$50 \log_{10} [U_B]$, dB	73.0	73.0	73.0	73.0	73.0	73.0	73.0	73.0
$10 \log_{10} [S_B]$, dB	−1.6	−1.6	−1.6	−1.6	−1.6	−1.6	−1.6	−1.6
$10 \log_{10} [D_B]$, dB	−0.3	−0.3	−0.3	−0.3	−0.3	−0.3	−0.3	−0.3
Δr, dB	6.2	5.7	5.1	4.5	4.0	3.4	2.9	2.3
ΔT, dB	0.0	0.0	0.0	0.0	0.0	0.0	0.0	0.0
$L_W(f_o)_b$, dB	73.2	70.6	67.4	63.4	58.6	53.1	46.9	39.9
$20 \log_{10} (D_M/D_B)$, dB	6.2	6.2	6.2	6.2	6.2	6.2	6.2	6.2
	3.0	3.0	3.0	3.0	3.0	3.0	3.0	3.0
$L_W(f_o)_m$, dB	82.4	79.8	76.6	72.6	67.8	62.3	56.1	49.1

EXAMPLE 4.4: T–Junction

Determine the regenerated sound power levels associated with a T–junction that exist in the branch and main ducts given the following information:

Main Duct: Rectangular — 12 in. by 36 in., volume flow rate — 12,000 cfm
Branch Duct: Rectangular — 12 in. by 18 in., volume flow rate — 6,000 cfm
Radius of bend or elbow: 0.0 in.
No dampers, elbows, or branch takeoffs are within five main duct diameters of the junction.

SOLUTION

Step 1: Determine the values of D_B and D_M.

$$D = \left[\frac{4 \cdot 12 \cdot 36}{\pi \cdot 144}\right]^{1/2} = 1.95 \text{ ft.} \qquad D = \left[\frac{4 \cdot 12 \cdot 18}{\pi \cdot 144}\right]^{1/2} = 1.38 \text{ ft.}$$

Step 2: Determine the values of U_B and U_M.

$$U_M = \frac{12000 \cdot 144}{12 \cdot 36 \cdot 60} = 66.67 \text{ ft/s} . \qquad U_B = \frac{6000 \cdot 144}{12 \cdot 18 \cdot 60} = 66.67 \text{ ft/s} .$$

Step 3: Determine the ratios, D_m/D_B and m.

$$\frac{D_M}{D_B} = \frac{1.95}{1.38} = 1.41 . \qquad m = \frac{66.67}{66.67} = 1.00 .$$

Step 4: Determine the rounding parameter, RD.

$$RD = \frac{0}{12 \cdot 1.38} = 0 .$$

The results are tabulated below.

	1/1 Octave Band Center Frequency – Hz							
	63	125	250	500	1000	2000	4000	8000
S_t	1.3	2.6	5.2	10.4	20.7	41.5	82.9	165.8
K_J, dB)	−11.1	−16.9	−23.6	−31.2	−49.7	−49.1	−59.4	−70.7
$10 \log_{10}[f_o/63]$, dB	0.0	3.0	6.0	9.0	12.0	15.0	18.0	21.0
$50 \log_{10}[U_B]$, dB	91.2	91.2	91.2	91.2	91.2	91.2	91.2	91.2
$10 \log_{10}[S_B]$, dB	1.8	1.8	1.8	1.8	1.8	1.8	1.8	1.8
$10 \log_{10}[D_B]$, dB	1.4	1.4	1.4	1.4	1.4	1.4	1.4	1.4
Δr, dB	6.6	6.0	5.5	4.9	4.3	3.8	3.2	2.7
ΔT, dB	0.0	0.0	0.0	0.0	0.0	0.0	0.0	0.0
$L_W(f_o)_b$, dB	89.8	86.5	82.3	77.1	71.1	64.1	56.2	47.4
	3.0	3.0	3.0	3.0	3.0	3.0	3.0	3.0
$L_W(f_o)_m$, dB	92.8	89.5	85.3	80.1	74.1	67.1	59.2	50.4

EXAMPLE 4.5: 90° Elbow without Turning Vanes

Determine the regenerated sound power levels associated with a 90° elbow without turning vanes given the following information:

Main Duct: Rectangular – 12 in. by 36 in., volume flow rate – 12,000 cfm
Branch Duct: Rectangular – 12 in. by 36 in., volume flow rate – 12,000 cfm
Radius of bend or elbow: – 0.0 in.

No dampers, elbows, or branch takeoffs are within five main duct diameters of the junction.

SOLUTION

Step 1: Determine the values of D_B and D_M.

$$D_M = \left[\frac{4 \cdot 12 \cdot 36}{\pi \cdot 144}\right]^{1/2} = 1.95 \text{ ft.} \qquad D_B = \left[\frac{4 \cdot 12 \cdot 36}{\pi \cdot 144}\right] = 1.95 \text{ ft.}$$

Step 2: Determine the values of U_B and U_M.

$$U_M = \frac{12000 \cdot 144}{12 \cdot 36 \cdot 60} = 66.67 \text{ ft/s}. \qquad U_B = \frac{12000 \cdot 144}{12 \cdot 36 \cdot 60} = 66.67 \text{ ft/s}.$$

Step 3: Determine the ratios, D_m/D_B and m.

$$\frac{D_M}{D_B} = \frac{1.95}{1.95} = 1.00. \qquad m = \frac{66.67}{66.67} = 1.00.$$

Step 4: Determine the rounding parameter, RD.

$$RD = \frac{0}{12 \cdot 1.95} = 0.$$

The results are tabulated below.

	1/1 Octave Band Center Frequency — Hz
	63	125	250	500	1000	2000	4000	8000
S_t	1.8	3.7	7.3	14.7	29.3	58.6	117.3	234.8
K_J, dB	−13.9	−20.1	−27.2	−35.3	−44.3	−54.1	−64.9	−76.7
$10 \log_{10} [f_o/63]$, dB	0.0	3.0	6.0	9.0	12.0	15.0	18.0	21.0
$50 \log_{10} [U_B]$, dB	91.2	91.2	91.2	91.2	91.2	91.2	91.2	91.2
$10 \log_{10} [S_B]$, dB	4.8	4.8	4.8	4.8	4.8	4.8	4.8	4.8
$10 \log_{10} [D_B]$, dB	2.9	2.9	2.9	2.9	2.9	2.9	2.9	2.9
Δr, dB	6.3	5.7	5.2	4.6	4.1	3.5	2.9	2.4
ΔT, dB	0.0	0.0	0.0	0.0	0.0	0.0	0.0	0.0
$L_W(f_o)_b$, dB	91.3	87.5	82.5	77.2	70.7	63.3	54.9	45.6
	0.0	0.0	0.0	0.0	0.0	0.0	0.0	0.0
$L_W(f_o)_m$, dB	91.3	87.5	82.5	77.2	70.7	63.3	54.9	45.6

EXAMPLE 4.6: 90° Branch Takeoff

Determine the regenerated sound power levels associated with a 90° branch takeoff that exist in the branch and main ducts given the following information:

Main Duct: Rectangular — 12 in. by 36 in., volume flow rate — 12,000 cfm
Branch Duct: Rectangular — 10 in. by 10 in., volume flow rate — 1,200 cfm
Radius of bend or elbow: 0.0 in.
No dampers, elbows, or branch takeoffs are within five main duct diameters of the junction.

SOLUTION

Step 1: Determine the values of D_B and D_M.

$$D_M = \left[\frac{4 \cdot 12 \cdot 36}{\pi \cdot 144}\right]^{1/2} = 1.95 \text{ ft.} \qquad D_B = \left[\frac{4 \cdot 10 \cdot 10}{\pi \cdot 144}\right]^{1/2} = 0.94 \text{ ft.}$$

Step 2: Determine the values of U_B and U_M.

$$U_M = \frac{12000 \cdot 144}{12 \cdot 36 \cdot 60} = 66.67 \text{ ft/s} . \qquad U_B = \frac{1200 \cdot 144}{10 \cdot 10 \cdot 60} = 28.80 \text{ ft/s} .$$

Step 3: Determine the ratios, D_m/D_B and m.

$$\frac{D_M}{D_B} = \frac{1.95}{0.94} = 2.06 . \qquad m = \frac{66.67}{28.80} = 2.31 .$$

Step 4: Determine the rounding parameter, RD.

$$RD = \frac{0}{12 \cdot 0.94} = 0 .$$

The results are tabulated below.

	\multicolumn{8}{c}{1/1 Octave Band Center Frequency — Hz}
	63	125	250	500	1000	2000	4000	8000
S_t	2.0	4.1	8.2	16.3	32.6	65.3	130.6	261.2
K_J, dB		-4.2	-9.1	-14.9	-21.3	-28.5	-36.4	-45.1	-54.5
$10 \log_{10} [f_o/63]$, dB	0.0	3.0	6.0	9.0	12.0	15.0	18.0	21.0
$50 \log_{10} [U_B]$, dB	73.0	73.0	73.0	73.0	73.0	73.0	73.0	73.0
$10 \log_{10} [S_B]$, dB	-1.6	-1.6	-1.6	-1.6	-1.6	-1.6	-1.6	-1.6
$10 \log_{10} [D_B]$, dB	-0.3	-0.3	-0.3	-0.3	-0.3	-0.3	-0.3	-0.3
Δr, dB	6.2	5.7	5.1	4.5	4.0	3.4	2.9	2.3
ΔT, dB	0.0	0.0	0.0	0.0	0.0	0.0	0.0	0.0
$L_W(f_o)_b$, dB	73.2	70.6	67.4	63.4	58.6	53.1	46.9	39.9
$20 \log_{10} [D_M/D_B]$	6.2	6.2	6.2	6.2	6.2	6.2	6.2	6.2
$L_W(f_o)_m$, dB	79.4	76.8	73.6	69.6	64.8	59.3	53.1	46.1

4.4 DIFFUSERS

Diffusers generate and radiate high–frequency sound into a space. Whenever possible, it is desirable to use manufacturers' supplied data. Manufacturers' data should be obtained according to ARI Standard 880–87, "Industry Standard for Air Terminals" (51). However, when such data are not available, the 1/1 octave band sound power levels for selected types of diffusers can be estimated by the procedure described below.

Sound radiation associated with air flow through diffusers and diffusers with porous plates that terminate air conditioning ducts is similar to sound radiation associated with air flowing over a spoiler. The interaction of the airflow and diffuser guide vanes behaves as an acoustic dipole. Thus, the associated sound power is proportional to the sixth power of flow velocity and the third power of pressure. The pressure drop across a diffuser can be specified by the normalized pressure drop coefficient, ξ (3). ξ is given by

$$\xi = 334.9 \frac{\Delta P}{\rho \cdot u^2} \tag{4.27}$$

where ΔP is the pressure drop across a diffuser (in. H_2O), ρ is the density of air (lb_m/ft^3), and u is the mean flow speed (ft/s) of the air in the duct prior to the diffuser. For most situations, $\rho = 0.0749\ lb_m/ft^3$. u is obtained from

$$u = \frac{Q}{60 \cdot S} \tag{4.28}$$

where Q is the flow volume (ft^3/min) and S is the duct cross–sectional area (ft^2) prior to the diffuser. The overall sound power level, $L_{W(overall)}$ (dB), associated with a diffuser is given by

$$L_{W(overall)} = 10 \log_{10} [S] + 30 \log_{10} [\xi] + 60 \log_{10} [u] - 31.3 \tag{4.29}$$

where ξ, u, and S are as defined before (3).

The peak frequency, f_p (Hz), associated with sound generated by diffusers can be approximated by

$$f_p = 48.8\ u \tag{4.30}$$

where u is as defined above (3). The shape of the 1/1 octave band sound spectrum for a diffuser is similar to that shown in Figure 4.8. If the diffusers are generic rectangular, round, and square perforated face (with round inlet) diffusers, the equation for the curve in Figure 4.8 is given by

$$C = -5.82 - 0.15 \cdot A - 1.13 \cdot A^2 \tag{4.31}$$

DUCT ELEMENT REGENERATED SOUND POWER

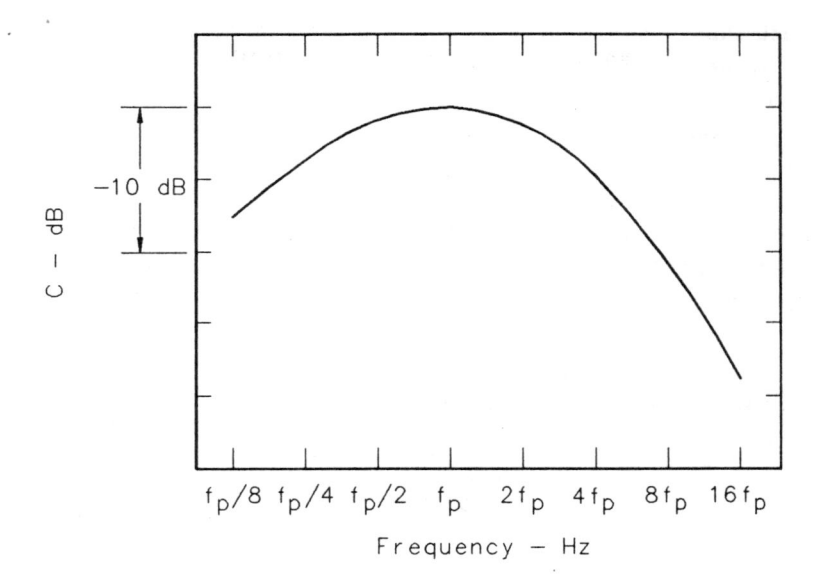

Figure 4.8 Generalized 1/1 Octave Band Spectrum Shape Associated with Diffuser Noise

for generic round diffusers and by

$$C = -11.82 - 0.15 \cdot A - 1.13 \cdot A^2 \tag{4.32}$$

for generic rectangular and square perforated face (with round inlet) diffusers, where

$$A = I - II; \tag{4.33}$$

I = 1 for 63 Hz, 2 for 125 Hz, 3 for 250 Hz, etc.; and II is dependent upon peak frequency and is specified by:

$$
\begin{aligned}
0 \leq f_p < 44 \text{ Hz} & \quad II = 0 \\
44 \leq f_p < 88 \text{ Hz} & \quad II = 1 \\
88 \leq f_p < 177 \text{ Hz} & \quad II = 2 \\
177 \leq f_p < 355 \text{ Hz} & \quad II = 3 \\
355 \leq f_p < 710 \text{ Hz} & \quad II = 4 \\
710 \leq f_p < 1420 \text{ Hz} & \quad II = 5 \\
1420 \leq f_p < 2840 \text{ Hz} & \quad II = 6 \\
2840 \leq f_p < 5680 \text{ Hz} & \quad II = 7 \\
5680 \leq f_p < 11360 \text{ Hz} & \quad II = 8 \, .
\end{aligned}
$$

Equation (4.32) can also be used for generic slot diffusers that do not have special plenum or damper systems. For rectangular slot diffusers, S and u in Equation (4.29) are the cross–sectional area and flow velocity just prior to the slots. The 1/1 octave band sound power levels associated with generic diffusers are given by

$$L_W = 10 \log_{10}[S] + 30 \log_{10}[\xi] + 60 \log_{10}[u] - 31.3 + C \, . \tag{4.34}$$

The sound power levels predicted by Equation (4.33) usually yield NC levels that are within five points of corresponding levels that are published by manufacturers when an 8 to 10 dB room correction is applied to each 1/1 octave band to convert from sound power levels to corresponding sound pressure levels.

The method for determining the sound power levels associated with generic diffusers described above does not apply for diffusers that have specially designed plenum and damper systems. When this is the case, the sound power levels of a diffuser can be estimated by using the manufacturer's published NC levels for a specified diffuser system and the related pressure drop, ΔP, and flow velocity, u, associated with the point of operation of the diffuser. The flow velocity, u, and corresponding peak frequency, f_p, are determined as described above. The curve in Figure 4.8 is shifted such that f_p corresponds to the 1/1 octave frequency band that contains f_p. Position the curve so that it is tangent to the NC curve that corresponds to the NC level published by the manufacturer for the specified point of operation. Read the related 1/1 octave band sound pressure levels. Finally, add 10 dB to all of the 1/1 octave band sound pressure levels to obtain the 1/1 octave band sound power levels of the diffuser.

EXAMPLE 4.7

A rectangular diffuser has the following duct dimensions prior to the diffuser: 12 in by 16 in. The volume flow rate is $Q = 1200$ ft^3/min, and the pressure drop across the diffuser is $\Delta P = 0.3$ in. H$_2$O. Determine the 1/1 octave band sound power levels associated with the diffuser.

SOLUTION

The cross–sectional area, S, and flow velocity, u, are

$$S = \frac{12 \cdot 16}{144} = 1.33 \text{ ft}^2 \ . \qquad u = \frac{1,200}{60 \cdot 1.33} = 15 \text{ ft/s} \ .$$

The normalized pressure drop coefficient, ξ, is

$$\xi = \frac{334.9 \cdot 0.3}{0.0749 \cdot 15^2} = 5.96 \ .$$

The overall sound power level is

$$L_{p(\text{overall})} = 10 \log_{10}[1.33] + 30 \log_{10}[5.96] + 60 \log_{10}[15] - 31.3 = 68.1 \text{ dB} \ .$$

The frequency, f_p, is

$$f_p = 48.8 \cdot 15 = 732 \text{ Hz} \ .$$

732 Hz is between 710 Hz and 1420 Hz. Thus, II = 5. The results are tabulated below.

	\multicolumn{8}{c}{1/1 Octave Band Center Frequency – Hz}
	63	125	250	500	1000	2000	4000	8000
$L_{W(\text{overall})}$, dB	63.8	63.8	63.8	63.8	63.8	63.8	63.8	63.8
A	−4	−3	−2	−1	0	1	2	3
C, dB (eq. 4.32)	−29.3	−21.5	−16.0	−12.8	−11.8	−13.1	−16.6	−22.4
L_W, dB	34.5	42.3	47.8	51.0	52.0	50.7	47.2	41.4

DUCT ELEMENT SOUND ATTENUATION

CHAPTER 5

5.1 PLENUM CHAMBERS

Plenum chambers are often used to "smooth out" turbulent air flow associated with air as it leaves the outlet section of a fan and before it enters the ducted air distribution system of a building. The plenum chamber is usually placed between the discharge section of a fan and the main duct of the air distribution system. These chambers are usually lined with acoustically absorbent material to reduce fan and other types of noise. Plenum chambers are usually large rectangular enclosures with an inlet and one or more outlet sections. Work originally presented by Wells (48) and later by Beranek (3) and Reynolds (27) indicates that the transmission loss associated with a plenum chamber can be expressed as

$$TL = -10 \log_{10} \left[S_{out} \left[\frac{Q \cdot \cos\theta}{4\pi \cdot r^2} + \frac{1-\alpha}{S \cdot \overline{\alpha}} \right] \right]. \qquad (5.1)$$

Referring to Figure 4.1, S_{out} is the area (ft^2) of the output section of the plenum, S is the total inside surface area (ft^2) of the plenum minus the inlet and outlet areas, r is the distance (ft) between the centers of the inlet and outlet sections of the plenum, and $\overline{\alpha}$ is the average absorption coefficient of the plenum lining. $\overline{\alpha}$ is given by

$$\overline{\alpha} = \frac{S_1 \alpha_1 + S_2 \alpha_2}{S} \qquad (5.2)$$

where α_1 and S_1 are the sound absorption coefficient and corresponding surface area (ft^2) of any bare or unlined inside surfaces of the plenum chamber, and α_2 and S_2 are the sound absorption coefficient and corresponding surface area (ft^2) of the acoustically lined inside surfaces of the plenum chamber. Often, one hundred percent of the inside surfaces of a plenum chamber are lined with a sound–absorbing material. For these situations, $\overline{\alpha} = \alpha_2$.

Q in Equation (5.1) is the directivity factor, which equals 2 if the inlet section is near the center of the side of the plenum on which it is located. This corresponds to the situation where sound from the inlet section of the plenum chamber is radiating into half space. Q equals 4 if the inlet section is located in the corner where two sides of the plenum come together. This corresponds to the situation where sound from the inlet section is radiating into quarter space.

θ in Equation (5.1) is the angle of the vector representing r relative to the horizontal plane. $\cos\theta$ and r can be written as

$$r = \sqrt{rh^2 + rv^2} \qquad (5.3)$$

$$\cos\theta = \frac{rh}{r} \qquad (5.4)$$

where rh and rv are the horizontal and vertical distances (ft), respectively, between the inlet and outlet sections of the plenum (Figure 5.1).

Equation (5.1) treats a plenum as if it were a large enclosure. Thus, Equation (5.1) is valid only for the case where the wavelength of sound is small compared to the characteristic dimensions of the plenum (3,27,48). For frequencies that correspond to plane wave propagation in the duct, the results predicted by Equation (5.1) are usually not valid. Plane wave propagation in a duct exists at frequencies below

$$f_{co} = \frac{c_o}{2a} \qquad (5.5)$$

where c_o is the speed of sound in air (ft/s) and a is the larger cross–sectional dimension (ft) of a rectangular duct, or below

$$f_{co} = 0.586 \frac{c_o}{d} \qquad (5.6)$$

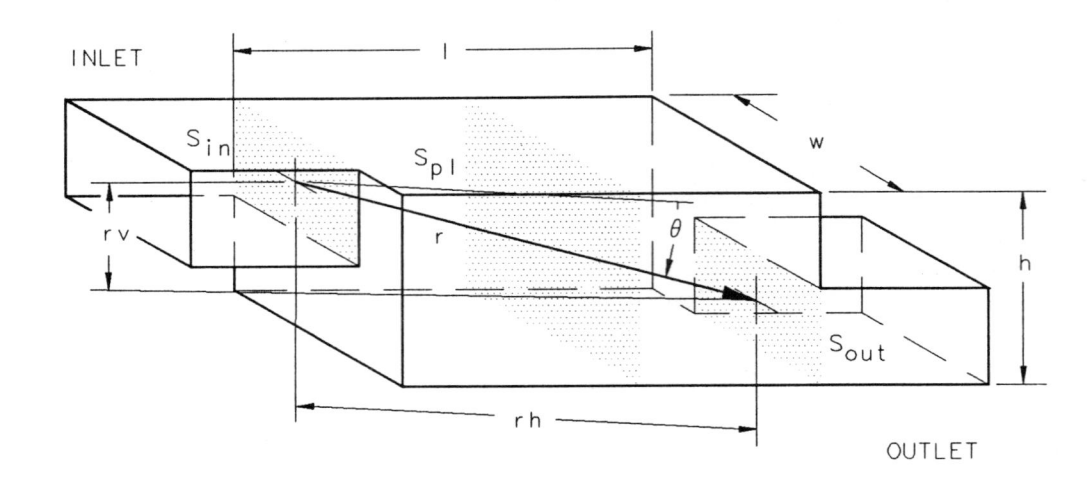

Figure 5.1 Schematic of a Plenum Chamber

where d is the diameter (ft) of a circular duct. The cutoff frequency, f_{co}, is the frequency above which plane waves no longer propagate in a duct. At these higher frequencies, the waves that propagate in the duct are referred to as cross or spinning modes (3,27). At frequencies below f_{co}, the plenum chamber can be treated as an acoustically lined expansion chamber. The equation for the transmission loss of an acoustically lined expansion chamber is

$$TL = 10 \log_{10}\left[\left[\cosh\left[\frac{\sigma l}{2}\right] + \frac{1}{2}\left[m + \frac{1}{m}\right]\sinh\left[\frac{\sigma l}{2}\right]\right]^2 \cdot \cos^2\left[\frac{2\pi \cdot f \cdot l}{c_o}\right]\right.$$
$$\left. + \left[\sinh\left[\frac{\sigma l}{2}\right] + \frac{1}{2}\left[m + \frac{1}{m}\right]\cosh\left[\frac{\sigma l}{2}\right]\right]^2 \cdot \sin^2\left[\frac{2\pi \cdot f \cdot l}{c_o}\right]\right] \tag{5.7}$$

where σ is sound attenuation per unit length in the chamber (dB/ft), l is the horizontal length of the plenum chamber (ft), c_o is the speed of sound in air (ft/s), f is frequency (Hz), and m is the ratio of the cross–sectional area of the plenum divided by the cross–sectional area of the inlet section of the plenum (3). m is given by

$$m = \frac{S_{pl}}{S_{in}} \quad \text{(refer to Figure 5.1)}. \tag{5.8}$$

For frequencies less than f_{co}, the transmission loss of a plenum is given by Equation (5.7). For frequencies greater than or equal to f_{co}, the transmission loss of a plenum is given by Equation (5.1). f_{co} associated with Equations (5.5) and (5.6) is calculated on the basis of the inlet section of the plenum. Table 5.1 gives the absorption coefficients of typical plenum materials (3,25,27).

Equations for σl for the 1/1 octave frequency bands from 63 Hz to 500 Hz are:

63 Hz:	$\sigma l = 0.00306 \cdot (P/A)^{1.959} \cdot t^{0.917} \cdot l$	(5.9)
125 Hz:	$\sigma l = 0.01323 \cdot (P/A)^{1.410} \cdot t^{0.941} \cdot l$	(5.10)
250 Hz:	$\sigma l = 0.06244 \cdot (P/A)^{0.824} \cdot t^{1.079} \cdot l$	(5.11)
500 Hz:	$\sigma l = 0.23380 \cdot (P/A)^{0.500} \cdot t^{1.087} \cdot l$	(5.12)

where P/A is the perimeter (P) of the cross section of the plenum chamber (ft) divided by the area (A or S_{pl}) of the cross section of the plenum chamber (ft²), t is the thickness of the fiberglass insulation (in.) used to line the

DUCT ELEMENT SOUND ATTENUATION

Table 5.1 Absorption Coefficients for Selected Plenum Materials

	\multicolumn{7}{c}{1/1 Octave Band Center Frequency — Hz}						
	63	125	250	500	1000	2000	4000
Non–Sound–Absorbing Materials							
Concrete	0.01	0.01	0.01	0.02	0.02	0.02	0.03
Bare Sheet Metal	0.04	0.04	0.04	0.05	0.05	0.05	0.07
Sound–Absorbing Materials							
1 in. 3.0 lb/ft^3 Fiberglass Insulation Board	0.02	0.03	0.22	0.69	0.91	0.96	0.99
2 in. 3.0 lb/ft^3 Fiberglass Insulation Board	0.18	0.22	0.82	1.00	1.00	1.00	1.00
3 in. 3.0 lb/ft^3 Fiberglass Insulation Board	0.48	0.53	1.00	1.00	1.00	1.00	1.00
4 in. 3.0 lb/ft^3 Fiberglass Insulation Board	0.76	0.84	1.00	1.00	1.00	1.00	0.97

inside surfaces of the plenum, and l is the length (ft) of the plenum chamber (23,41). Equation (5.1) will nearly always apply at frequencies of 1,000 Hz and above.

EXAMPLE 5.1

A plenum chamber is 6 ft high, 4 ft wide, and 10 ft long. The configuration of the plenum is similar to that shown in Figure 5.1. The inlet is 36 in. wide by 24 in. high. The outlet is 36 in. wide by 24 in. high. The horizontal distance between centers of the plenum inlet and outlet is 10 ft. The vertical distance is 4 ft. The plenum is lined with 1–in.–thick, 3.0 lb/ft^3 density fiberglass insulation board. One hundred percent of the inside surfaces of the plenum are lined with the fiberglass insulation. Determine the transmission loss associated with this plenum. For this example, assume Q = 4.

SOLUTION

The areas of the inlet section, outlet section, and plenum cross section are

$$S_{in} = \frac{24 \cdot 36}{144} = 6 \text{ ft}^2 \qquad S_{out} = \frac{24 \cdot 36}{144} = 6 \text{ ft}^2 \qquad S_{pl} = 4 \cdot 6 = 24 \text{ ft}^2.$$

The values of r and $\cos\theta$ are

$$r = \sqrt{10^2 + 4^2} = 10.77 \text{ ft} \qquad \cos\theta = \frac{10}{10.77} = 0.93.$$

The total inside surface area of the plenum is

$$S = 2 \cdot (4 \cdot 6) + 2 \cdot (4 \cdot 10) + 2 \cdot (6 \cdot 10) - 12 = 236 \text{ ft}^2.$$

The values of P/A, m, and f_{co} are

$$P/A = \frac{2 \cdot (4 + 6)}{4 \cdot 6} = 0.83 \qquad m = \frac{24}{6} = 4 \qquad f_{co} = \frac{1125}{2 \cdot 3} = 187.5 \text{ Hz}.$$

Thus, Equation (5.8) is used for the 63 Hz and 125 Hz 1/1 octave bands and Equation (5.1) is used for the 250 Hz through 4,000 Hz 1/1 octave bands. The results are tabulated below.

	1/1 Octave Band Center Frequency – Hz						
	63	125	250	500	1000	2000	4000
σl	0.093	0.444					
m	4	4					
$2\pi \cdot f \cdot l/c_o$	3.519	6.981					
$Q\cos\theta/4\pi r^2$ (x 10^3)			2.55	2.55	2.55	2.55	2.55
$(1-\overline{\alpha})/S\overline{\alpha}$ (x 10^3)			15.0	1.90	.0419	.0177	.00428
TL, dB	2.1	5.7	9.8	15.7	17.5	17.9	18.1

5.2 UNLINED RECTANGULAR SHEET METAL DUCTS

Straight unlined rectangular sheet metal ducts provide a small amount of sound attenuation. At low frequencies, the attenuation is significant and it tends to decrease as frequency increases. Work reported by Hal Sabine indicated that even though there was significant sound attenuation in rectangular ducts at low frequencies, it tended to be "irregular with respect to duct size" (39). Table 5.2 shows a compilation of data from Sabin, Ver, and Chaddock that was completed by Reynolds and Bledsoe. (10,14,29,45) The unit on P/A is 1/ft. A multi–variable regression analysis of the data in Table 5.2 for the 1/1 octave band center frequencies between 63 Hz and 250 Hz yielded

$$\text{ATTN} = 17.0 \cdot (P/A)^{-0.25} \cdot \text{FREQ}^{-0.85} \cdot L \quad \text{for } P/A \geq 3 \tag{5.13}$$

$$\text{ATTN} = 1.64 \cdot (P/A)^{0.73} \cdot \text{FREQ}^{-0.58} \cdot L \quad \text{for } P/A < 3 \tag{5.14}$$

where ATTN is the total attenuation (dB) in the unlined rectangular duct, P is the length of the duct perimeter (ft), A is the duct cross–sectional area (ft^2), FREQ is the 1/1 octave band center frequency (Hz), and L is the duct length (ft). An analysis by Reynolds and Bledsoe (29) of the sound attenuation data at frequencies above 250 Hz indicated that

$$\text{ATTN} = 0.02 \cdot (P/A)^{0.8} \cdot L. \tag{5.15}$$

Table 5.3 shows the tabulated results that correspond to Equations (5.13) through (5.15). If the rectangular duct is externally lined with fiberglass, multiply the results associated with Equation (5.13) or (5.14) by a factor of 2 (10,14,29,36,37,39,45).

The attenuations values shown in Tables 5.2 and 5.3 and the corresponding attenuation values predicted by Equations (5.13) through (5.15) apply only to rectangular sheet metal ducts that have gauges thicknesses that are selected according to SMACNA (Sheet Metal and Air Conditioning Contractors National Association) HVAC duct construction standards.

Table 5.2 Sound Attenuation in Unlined Rectangular Sheet Metal Ducts

P/A 1/ft.	Attenuation – dB/ft 1/1 Oct Band Center Freq – Hz		
	63	125	250
8.0	0.35	0.19	0.09
4.0	0.31	0.24	0.10
3.0	0.35	0.29	0.13
2.0	0.20	0.20	0.10
1.0	0.20	0.20	0.10
0.7	0.10	0.10	0.05

Table 5.3 Sound Attenuation in Unlined Rectangular Sheet Metal Ducts — Regression

		Attenuation — dB/ft			
		1/1 Octave Band Center Freq. — Hz			
Duct Size in. x in.	P/A 1/ft	63	125	250	Above 250
6 x 6	8.0	0.30	0.20	0.10	0.10
12 x 12	4.0	0.35	0.20	0.10	0.06
12 x 24	3.0	0.40	0.20	0.10	0.05
24 x 24	2.0	0.25	0.20	0.10	0.03
48 x 48	1.0	0.15	0.10	0.07	0.02
72 x 72	0.7	0.10	0.10	0.05	0.02

If duct is externally lined, multiply results associated with 63 Hz, 125 Hz and 250 Hz by 2.

EXAMPLE 5.2

A straight section of unlined rectangular duct has the following dimensions: height = 18 in., width = 12 in., and length = 20 ft. Determine the total sound attenuation in dB.

SOLUTION

$$P/A = \frac{2 \cdot (12+18) \cdot 12}{(12 \cdot 18)} = 3.333 \ 1/\text{ft}.$$

The tabulated results are shown below.

	1/1 Octave Band Center Frequency — Hz							
	63	125	250	500	1000	2000	4000	8000
Eq. (5.13), dB/ft	0.37	0.21	0.12					
Eq. (5.14), dB/ft				0.05	0.05	0.05	0.05	0.05
Duct length, ft	x20	x20	x20	x20	x20	x20	x20	x20
Total Atten., dB	7.4	4.2	2.4	1.0	1.0	1.0	1.0	1.0

5.3 ACOUSTICALLY LINED RECTANGULAR SHEET METAL DUCTS

Fiberglass internal duct lining for rectangular sheet metal ducts can be used to attenuate sound in ducts and to thermally insulate ducts. The thickness of duct linings associated with thermal insulation usually vary from 0.5 in. to 2.0 in. For fiberglass duct lining to be effective for attenuating sound, it must have a minimum thickness of 1.0 in.

Reynolds and Bledsoe conducted a multi–variable regression analysis of insertion loss data for acoustically lined rectangular sheet metal ducts that was obtained by Kuntz, Hoover and Kuntz, and Machen and Haines (22,23,24,29). The P/A values in the unit of 1/ft of ducts tested ranged from 1.1667 to 6; the thickness of the fiberglass duct lining was either 1 in. or 2 in.; and the density of the fiberglass duct lining ranged from 1.5 to 3.0 lb/ft^3. The regression equation for insertion loss in acoustically lined rectangular ducts is

$$IL = B \cdot (P/A)^C \cdot t^D \cdot L \ (\text{dB}) \tag{5.16}$$

where IL is the insertion loss (dB); P/A is the perimeter divided by the cross–sectional area of the free area inside the duct (1/ft); B, C, and D are regression constants that are a function of the 1/1 octave band center frequency; t is lining thickness (in.); and L is duct length (ft). The values for B, C, and D are given in Table 5.4 for 1/1 octave band center frequencies from 63 Hz to 8,000 Hz.

The following observations are made relative to the regression analysis reported by Reynolds and Bledsoe (29). Caution must be exercised when extrapolating the values of insertion loss beyond the range of the parameters

Table 5.4 Constants for Use in Equation (5.16)

1/1 Octave Band Center Freq. — Hz	B	C	D
63	0.0133	1.959	0.917
125	0.0574	1.410	0.941
250	0.271	0.824	1.079
500	1.0147	0.500	1.087
1,000	1.770	0.695	0.0
2,000	1.392	0.802	0.0
4,000	1.518	0.451	0.0
8,000	1.581	0.219	0.0

associated with the data used to obtain Equation (5.16). The insertion loss values predicted by Equation (5.16) are valid only for 1/1 octave frequency bands. The regression analyses indicated that for the samples tested, the insertion loss of acoustically lined rectangular sheet metal ducts is not a function of the density of the fiberglass lining when the density of the material is between 1.5 and 3.0 lb/ft^3. At 1/1 octave band center frequencies of 1,000 Hz and above, the insertion loss is not a function of lining thickness.

The insertion loss described by Equation (5.16) is the difference in the sound pressure level measured in a reverberation chamber with sound propagating through an unlined section of rectangular duct minus the corresponding sound pressure level that is measured when the unlined section of rectangular duct is replaced with a similar section of acoustically lined rectangular duct. As mentioned in the section on unlined rectangular ducts, the sound attenuation associated with unlined rectangular duct can be significant at low frequencies. This attenuation is, in effect, subtracted out during the process of calculating the insertion loss from measured data. Even though it is not known for certain at this time, it is believed that this attenuation should be added to the insertion loss of correspondingly sized acoustically lined rectangular ducts to obtain the total sound attenuation of acoustically lined rectangular ducts. The sound attenuation, ATTN, in unlined rectangular ducts for the 1/1 octave band center frequencies from 63 Hz to 250 Hz is given by Equations (5.13) and (5.14). For 1/1 octave band center frequencies above 250 Hz, the sound attenuation, ATTN, is given by Equation (5.15). The total sound attenuation, ATTN(T), in acoustically lined rectangular ducts is obtained from

$$ATTN(T) = ATTN + IL . \tag{5.17}$$

Because of structure—borne sound that is transmitted in and through the duct wall, the total sound attenuation in lined rectangular sheet metal ducts usually does not exceed 40 dB. Thus, the maximum allowable sound attenuation in Equation (5.17) is 40 dB. Insertion loss and attenuation values obtained from equations (5.16) and (5.17) apply only to rectangular sheet metal ducts that have gauge thicknesses that are selected according to SMACNA HVAC duct construction standards.

EXAMPLE 5.3

A straight section of acoustically lined rectangular duct has the following free inside dimensions: height = 24 in., width = 36 in., and length = 10 ft. The duct is lined with 1-in.-thick, 1.5 lb/ft^3 fiberglass duct liner. Determine the total sound attenuation in the 10-ft section of acoustically lined rectangular duct.

SOLUTION

$$P/A = \frac{2 \cdot (24+36) \cdot 12}{24 \cdot 36} = 1.667 \ 1/\text{ft}.$$

The results are tabulated below.

	\multicolumn{8}{c}{1/1 Octave Band Center Frequency — Hz}
	63	125	250	500	1000	2000	4000	8000
Eq. (5.16), dB	0.4	1.2	4.1	13.1	25.2	21.0	19.1	17.7
Eq. (5.14), dB	2.2	1.4	0.9					
Eq. (5.15), dB				0.3	0.3	0.3	0.3	0.3
ATTN(T), dB	2.6	2.6	5.0	13.4	25.5	21.3	19.4	18.0

5.4 UNLINED CIRCULAR SHEET METAL DUCTS

As with unlined rectangular ducts, unlined circular ducts provide some sound attenuation that should be taken into account when designing a duct system. In contrast with rectangular ducts, circular ducts are much more rigid and, therefore, do not resonate or absorb as much sound energy. Because of this, circular ducts will only provide about one-tenth the sound attenuation at low frequencies as compared to the sound attenuation associated with rectangular ducts. Information in the chapter on sound and vibration in the ASHRAE 1987 HVAC Systems and Applications Handbook indicates:

"The natural (sound) attenuation for round ducts with or without external thermal insulation is about 0.03 dB/ft (0.1 dB/m) below 1,000 Hz, rising irregularly to 0.1 dB/ft (0.3 dB/m) at high frequencies." (36)

Information in the Woods Design for Sound manual is more specific than the information given in the ASHRAE handbook (14). This information is listed in Table 5.5. The sound attenuation information in the Woods manual is for the 1/1 octave band center frequencies from 125 Hz to 4,000 Hz. The information listed in Table 5.5 is extended to 63 Hz on the low end and to 4,000 Hz on the high end of the frequency range.

Table 5.5 Sound Attenuation in Straight Circular Ducts

	\multicolumn{7}{c}{Attenuation — dB/ft}
	\multicolumn{7}{c}{1/1 Octave Band Center Frequency — Hz}
Width — in.	63	125	250	500	1000	2000	4000
D ≤ 7	0.03	0.03	0.05	0.05	0.10	0.10	0.10
7 < D ≤ 15	0.03	0.03	0.03	0.05	0.07	0.07	0.07
15 < D ≤ 30	0.02	0.02	0.02	0.03	0.05	0.05	0.05
30 < D ≤ 60	0.01	0.01	0.01	0.02	0.02	0.02	0.02

EXAMPLE 5.4

A straight unlined circular duct has the following dimensions: diameter = 12 in. and length = 20 ft. Determine the total attenuation in dB.

SOLUTION

	\multicolumn{7}{c}{1/1 Octave Band Center Frequency, Hz}
	63	125	250	500	1000	2000	4000
Table 5.5, dB/ft	0.03	0.03	0.03	0.05	0.07	0.07	0.07
Length, ft	x20	x20	x20	x20	x20	x20	x20
Total Atten., dB	0.6	0.6	0.6	1.0	1.4	1.4	1.4

5.5 ACOUSTICALLY LINED CIRCULAR SHEET METAL DUCTS

Few data are available in the literature regarding the insertion loss of acoustically lined circular ducts. The data that are available are usually manufacturer's product data. The data that were used to develop an equation for determining the insertion loss of acoustically lined circular ducts were recently obtained by Bodley (8). The data were obtained for spiral, dual—wall circular ducts. The inside diameter of the ducts tested ranged from 6 to 60 in. The acoustical lining was a 0.75 lb/ft^3 density fiberglass blanket which ranged in thickness from 1 to 3 in. The fiberglass was covered with an internal liner of perforated galvanized steel that had an open area of 25%. Reynolds and Bledsoe conducted a multi—variable regression analysis on the data to determine the relationship between insertion loss values and the duct diameter and lining thickness for each of the 1/1 octave frequency bands between 63 Hz and 8,000 Hz (28). This analysis yielded

$$IL = (A + B \cdot t + C \cdot t^2 + D \cdot d + E \cdot d^2 + F \cdot d^3) \cdot L \tag{5.18}$$

where IL is insertion loss (dB), t is the lining thickness (in), d is the inside duct diameter (in), and L is the duct length (ft). The coefficients for Equation (5.18) for each of the 1/1 octave frequency bands are given in Table 5.6. At frequencies between 63 Hz and 500 Hz, the insertion loss is a function of both duct diameter and lining thickness. Above 1,000 Hz, the insertion loss is a function of duct diameter only. Because Equation (5.18) is a regression equation, it should not be extrapolated beyond the limits of the data on which the it is based. The sound attenuation of unlined circular ducts is negligible. Thus, it is not necessary to include it when calculating the total sound attenuation of lined circular ducts. Because of structure—borne sound that is transmitted through the duct wall, the total sound attenuation of lined circular ducts usually does not exceed 40 dB.

Table 5.6 Constants for Use in Equation (5.18)

1/1 Octave Band Center Freq.—Hz	A	B	C	D	E	F
63	0.2825	0.3447	−5.251E−2	−0.03837	9.1315E−4	−8.294E−6
125	0.5237	0.2234	−4.936E−3	−0.02724	3.377E−4	−2.49E−4
250	0.3652	0.7900	−0.1157	−1.834E−2	−1.211E−4	2.681E−4
500	0.1333	1.845	−0.3735	−1.293E−2	8.624E−5	−4.986E−6
1,000	1.933	0.0	0.0	6.135E−2	−3.891E−3	3.934E−5
2,000	2.730	0.0	0.0	−7.341E−2	4.428E−4	1.006E−6
4,000	2.800	0.0	0.0	−0.1467	3.404E−3	−2.851E−5
8,000	1.545	0.0	0.0	−5.452E−2	1.290E−3	−1.318E−5

EXAMPLE 5.5

Determine the sound attenuation in dB through a circular duct that has an inside diameter of 24 in. and a 1—in.—thick fiberglass lining. Assume the duct lining has a density of 0.75 lb/ft^3. The fiberglass lining is covered with an internal perforated galvanized steel liner that has an open area of 25%. The duct is 10 ft long.

SOLUTION

The insertion loss is calculated using Equation (5.18) and the corresponding coefficients in Table 5.6. For example, for the 63 Hz 1/1 octave band, Equation (5.18) will have the following form:

$$IL = (0.2825 + 0.3447 \cdot t - 5.251 \times 10^{-2} \cdot t^2 - 0.03837 \cdot d + 9.1331 \times 10^{-4} \cdot d^2 - 8.294 \times 10^{-6} \cdot d^3) \cdot L$$

where t is 1 in., d is 24 in., and L is 10 ft. Substituting in the values for t and d and reducing yields

$$IL = 0.065 \quad dB/ft.$$

The results are tabulated below.

	\multicolumn{8}{c}{1/1 Octave Band Center Frequency — Hz}
	63	125	250	500	1000	2000	4000	8000
IL, dB/ft	0.065	0.25	0.57	1.28	1.71	1.24	0.85	0.80
Length, ft	x10	x10	x10	x10	x10	x10	x10	x10
IL, dB	0.65	2.5	5.7	12.8	17.1	12.4	8.5	8.0

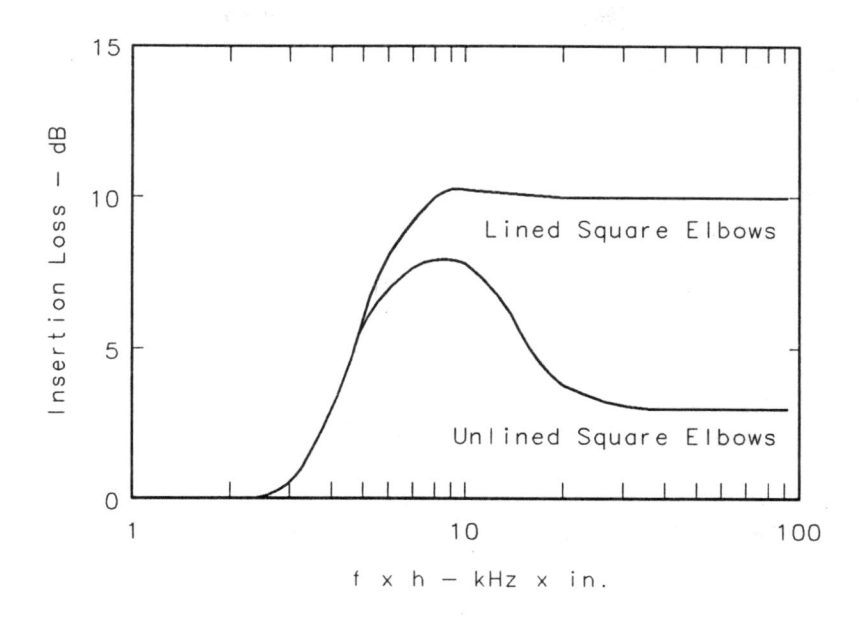

Figure 5.2 Insertion Loss Values for Unlined and Lined Square Elbows without Turning Vanes

5.6 RECTANGULAR SHEET METAL DUCT ELBOWS

The sound attenuation data for unlined and lined square elbows without turning vanes in the sound and vibration control chapters of the 1984 and earlier ASHRAE Systems Handbooks are referenced to curves that are contained in the book Noise Reduction by Beranek (4,38). These curves are shown in Figure 5.2. The original curves were plotted as a function of w/λ where w (Figure 5.3) is the duct dimension in the plane of the bend and λ is the wavelength that corresponds to 1/1 octave band center frequencies. For convenience, the curves in Figure 5.2 are plotted as a function of f x w where w (in.) is defined as before and f (kHz) is the 1/1 octave band center frequency. Modifications to the insertion loss values of lined and unlined bends have been proposed based on these curves. However, there have been no referenced or acceptable test results associated with these modifications. This also applies to the lined rectangular elbow insertion loss values given in the 1987 ASHRAE Handbook (36).

Table 5.7 displays insertion loss values for unlined square elbows without turning vanes (4). Table 5.8 shows a tabulation of the insertion loss values for lined square elbows without turning vanes (4). For lined square elbows, the duct lining must extend at least two duct widths, w, beyond the elbow and the thickness of the total lining thickness should be at least 10% of the duct width, w (4,38). The insertion loss tables for lined square elbows contained in the sound and vibration control chapters of the 1984 and earlier ASHRAE Systems Handbooks displayed insertion loss values of elbows with lining before the elbow, with lining after the elbow, and with lining before and after the elbow (37,38,40). In earlier versions of the Systems Handbook, these values were referenced to Figure 5.2. There appears to be no experimental justification in the literature for all of these configurations. There is justification only for the insertion loss values shown in Table 5.8. Typically, a lined elbow will be located in a duct that is lined before and after the elbow. Thus, it is assumed that the values in Table 5.8 apply to this situation.

The only early references associated with the insertion loss of round elbows and square elbows with turning vanes are contained in the chapters on sound control in early versions of the ASHRAE Guide and Data Book (40). Table 5.9 gives the insertion loss values associated with round elbows and square elbows with turning vanes. It appears that the insertion loss values in Table 5.9 were obtained from Figure 5.2 by smoothing out the "hump" in the curve associated with unlined square elbows. Later chapters on sound and vibration control in the ASHRAE Handbooks indicated that the insertion loss of unlined and lined square elbows with turning vanes should be

Square Elbow

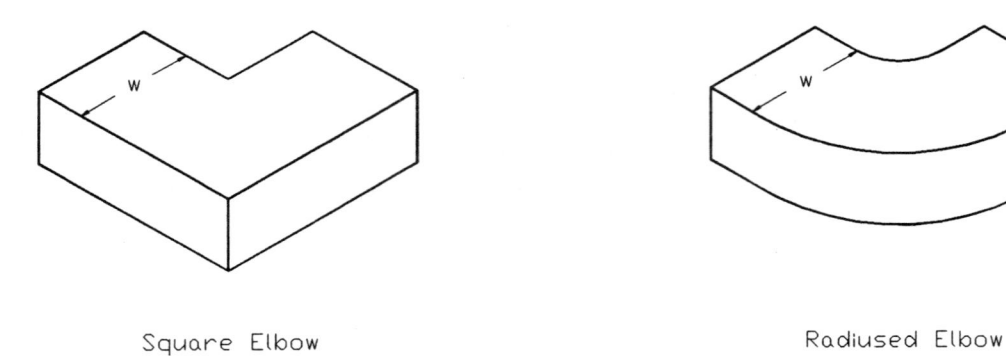

Radiused Elbow

Figure 5.3 Rectangular Duct Elbows

Table 5.7 Insertion Loss of Unlined Square Elbows without Turning Vanes

Duct Width inches	Insertion Loss — dB 1/1 Octave Band Center Frequency — Hz						
	63	125	250	500	1000	2000	4000
5	0	0	0	1	5	8	4
10	0	0	1	5	8	4	3
20	0	1	5	8	4	3	3
40	1	5	8	4	3	3	3

Table 5.8 Insertion Loss of Lined Square Elbows without Turning Vanes

Duct Width inches	Insertion Loss — dB 1/1 Octave Band Center Frequency — Hz						
	63	125	250	500	1000	2000	4000
5	0	0	0	1	6	11	10
10	0	0	1	6	11	10	10
20	0	1	6	11	10	10	10
40	1	6	11	10	10	10	10

Table 5.9 Insertion Loss of Unlined Round Elbows

Duct Diameter or Duct Width inches	Insertion Loss — dB 1/1 Octave Band Center Frequency — Hz						
	63	125	250	500	1000	2000	4000
5	0	0	0	1	2	3	3
10	0	0	1	2	3	3	3
20	0	1	2	3	3	3	3
40	1	2	3	3	3	3	3

be obtained by averaging the corresponding insertion loss values in Tables 5.7 and 5.9. Tables 5.10 and 5.11 display these values. The insertion loss values given in Tables 5.14 and 5.15 are only estimates. There are no experimental data to support these values.

Table 5.10 Insertion Loss of Unlined Square Elbows with Turning Vanes

Duct Width inches	Insertion Loss — dB 1/1 Octave Band Center Frequency — Hz						
	63	125	250	500	1000	2000	4000
5	0	0	0	1	4	6	4
10	0	0	1	4	6	4	3
20	0	1	4	6	4	3	3
40	1	4	6	4	3	3	3

Table 5.11 Insertion Loss of Lined Square Elbows with Turning Vanes

Duct Width inches	Insertion Loss — dB 1/1 Octave Band Center Frequency — Hz						
	63	125	250	500	1000	2000	4000
5	0	0	0	1	4	7	7
10	0	0	1	4	7	7	7
20	0	1	4	7	7	7	7
40	1	4	7	7	7	7	7

Table 5.12 Values of BW (kHz–in)

Duct Width inches	1/1 Octave Band Center Frequency — Hz						
	63	125	250	500	1000	2000	4000
5	—	—	1.25	2.5	5	10	20
10	—	1.25	2.5	5	10	20	40
20	1.25	2.5	5	10	20	40	80
40	2.5	5	10	20	40	80	160

Table 5.13 Insertion Loss of Unlined and Lined Square Elbows without Turning Vanes

BW	Insertion Loss – dB	
	Unlined Elbows	Lined Elbows
BW < 1.9	0	0
1.9 ≤ BW < 3.8	1	1
3.8 ≤ BW < 7.5	5	6
7.5 ≤ BW < 15	8	11
15 ≤ BW < 30	4	10
BW > 30	3	10

Table 5.14 Insertion Loss of Round Elbows

BW	Insertion Loss — dB
BW < 1.9	0
1.9 ≤ BW < 3.8	1
3.8 ≤ BW < 7.5	2
BW > 7.5	3

Table 5.15 Insertion Loss of Unlined and Lined Square Elbows with Turning Vanes

BW	Insertion Loss — dB Unlined Elbows	Lined Elbows
BW < 1.9	0	0
1.9 ≤ BW < 3.8	1	1
3.8 ≤ BW < 7.5	4	4
7.5 ≤ BW < 15	6	7
BW > 15	4	7

In order to use the above insertion loss values in an algorithm it was necessary to note that the insertion loss values are a function of f x w. Thus, the following relationship was used:

$$BW = f \cdot w \qquad (5.19)$$

where f is the 1/1 octave band center frequency (kHz) and w is the duct width or diameter (in.). This relationship gives the same diagonal symmetry that exists in Tables 5.7 through 5.11. This is illustrated in Table 5.12. Tables 5.13 through 5.15 show a recompilation of Tables 5.11 through 5.15, using Equation (5.19).

EXAMPLE 5.6

Determine the insertion loss (dB) of a 24-in. acoustically lined square elbow without turning vanes.

SOLUTION

	1/1 Octave Band Center Frequency — Hz							
	63	125	250	500	1000	2000	4000	8000
BW	1.5	3	6	12	24	48	96	192
IL (Table 5.13), dB	0	1	6	11	10	10	10	10

EXAMPLE 5.7

Determine the insertion loss (dB) of a round elbow constructed of a 12-in.-diameter unlined circular duct.

SOLUTION

	1/1 Octave Band Center Frequency — Hz							
	63	125	250	500	1000	2000	4000	8000
BW	0.76	1.5	3	6	12	24	48	96
IL (Table 5.14), dB	0	0	1	2	3	3	3	3

5.7 ACOUSTICALLY LINED CIRCULAR RADIUSED ELBOWS

Few data are available in the literature with regard to the insertion loss of acoustically lined radiused circular elbows. The data that are available are usually manufacturers' product data. The data that were used to develop an equation for determining the insertion loss of acoustically lined circular ducts were recently obtained by Bodley (8). The data were obtained for spiral dual wall circular ducts. The inside diameter of the elbows tested ranged from 6 to 60 in. The acoustical lining was a 0.75 lb/ft^3 density fiberglass blanket, which ranged in thickness from 1 to 3 in. The fiberglass was covered with an internal liner of perforated galvanized steel that had an open area of 25%. Reynolds and Bledsoe conducted a multi-variable regression analysis on the data to determine the relationship between insertion loss and the duct diameter and lining thickness for each of the 1/1 octave frequency bands between 63 Hz and 8,000 Hz (28). For ducts where $6 \leq d \leq 18$ in.,

$$IL \cdot \left[\frac{d}{r}\right]^2 = 0.485 + 2.094 \cdot \log_{10} [f \cdot d] + 3.172 \cdot \left[\log_{10} [f \cdot d]\right]^2$$
$$- 1.578 \cdot \left[\log_{10} [f \cdot d]\right]^4 + 0.085 \cdot \left[\log_{10} [f \cdot d]\right]^7 \quad (5.19)$$

and for ducts where $18 < d \leq 60$ in.,

$$IL \cdot \left[\frac{d}{r}\right]^2 = -1.493 + 0.538 \cdot t + 1.406 \cdot \log_{10} [f \cdot d] + 2.779 \cdot \left[\log_{10} (f \cdot d)\right]^2$$
$$- 0.662 \cdot \left[\log_{10} [f \cdot d]\right]^4 + 0.016 \cdot \left[\log_{10} [f \cdot d]\right]^7 \quad (5.20)$$

where f is the 1/1 octave band center frequency (Hz), d is the duct diameter (in), r is the radius of the elbow to the center line of the duct (in.), and t is the thickness (in.) of the acoustical duct liner. Equations (5.19) and (5.20) are seventh order polynomials. Thus, the equations should not be extrapolated beyond the specified limits for each equation. If the value for $IL \cdot (d/r)^2$ is negative in either Equation (5.19) or (5.20), set the value equal to zero. The relation that existed between r, d, and t for the elbows that were tested is

$$r = 1.5 \cdot d + 3 \cdot t . \quad (5.21)$$

EXAMPLE 5.8

Determine the insertion loss (dB) of a 24-in.-diameter acoustically lined circular elbow with a lining thickness of 2 in.

SOLUTION

$r = 1.5 \cdot (24) + 3 \cdot (2) = 42$ in.

$\left[\frac{d}{r}\right]^2 = \left[\frac{24}{42}\right]^2 = 0.327 .$

For a diameter of 24 in., Equation 5.20 applies. Thus,

$$IL \cdot \left[\frac{d}{r}\right]^2 = -1.493 + 0.538 \cdot (2) + 1.406 \cdot \log_{10} [f \cdot 24] + 2.779 \cdot \left[\log_{10} [f \cdot 24]\right]^2$$
$$- 0.662 \cdot \left[\log_{10} [f \cdot 24]\right]^4 + 0.016 \cdot \left[\log_{10} [f \cdot 24]\right]^7 .$$

The results are tabulated below.

	\multicolumn{8}{c}{1/1 Octave Band Center Frequency – Hz}
	63	125	250	500	1000	2000	4000	8000
$IL \cdot \left[\frac{d}{r}\right]^2$	0.0	0.85	2.1	3.5	4.6	5.1	5.0	4.5
$\times (r/d)^2$	3.06	3.06	3.06	3.06	3.06	3.06	3.06	3.06
IL, dB	0.0	2.6	6.4	10.7	14.1	15.6	15.3	13.8

5.8 DUCT SILENCERS

Duct silencers are often used as a means to attenuate unwanted noise in heating, ventilating, and air-conditioning systems. When duct silencers are used, the following parameters should be considered:

Insertion Loss — The difference between two sound power levels when measured at the same point before and after the silencer is installed.

Airflow Regenerated Noise — The sound power level generated by air flowing through a silencer.

Static Pressure Drop

Forward or Reverse Flow — Silencers have different acoustic and aerodynamic characteristics for forward and reverse flow directions.

There are two basic types of HVAC duct silencers: active and dissipative. Active duct silencer systems are rather new. They are very effective in attenuating low-frequency, pure-tone noise in a duct. They are also effective in attenuating low-frequency, broad-band noise. Active duct silencers consist of a microprocessor, two microphones placed a specified distance apart in a duct, and a speaker that is placed between the microphones and which is mounted external to the duct but radiates sound into the duct (Figure 5.4a). The microphone closest to a sound source that generates objectionable low-frequency noise senses the noise. The microphone signal is processed by the microprocessor, which generates a signal that is out of phase with the objectionable noise and is transmitted to the speaker. The speaker signal interferes with the objectionable noise, effectively attenuating it. The second microphone downstream of the speaker senses the attenuated noise and sends a corresponding feedback signal to the microprocessor, so the speaker signal can be adjusted, if necessary. Active duct silencer systems have no components that are located within a duct. Thus, they can be used to attenuate objectionable noise without introducing a pressure loss or regenerated noise into a duct. At present, not enough application data are available to develop an active silencer performance prediction algorithm.

Dissipative silencers are effective in attenuating broad-band noise; however, they introduce a pressure drop and regenerated noise into a duct. These should always be examined when considering the use of a dissipative silencer. Dissipative silencers can have a rectangular or circular cross section (Figure 5.4b and c). Rectangular silencers usually are available in several different cross-section and dimensional configurations and in 3 ft, 5 ft, 7 ft, and 10 ft lengths. Rectangular silencers have parallel sound absorbing surfaces, which usually are perforated sheet metal surfaces that cover cavities filled with either fiberglass or mineral wool. Circular silencers come in several different open-face diameters and usually have lengths that are a function of the open-face diameter. All circular silencers have a center body similar to the one shown in Figure 5.4c. This body is a cylindrical body with a perforated sheet metal surface and filled with either fiberglass or mineral wool. The outside shell of a circular silencer can be of either single- or double-wall construction. For single-wall construction, the outside shell is a solid cylindrical sheet metal shell that has a diameter equal to the open-face diameter of the silencer. For double-wall construction, the outside shell consists of two concentric cylindrical sheet metal shells. The outer shell is solid sheet metal. The inner shell is perforated sheet metal and it has a diameter equal to the open-face diameter of the silencer. The space between the two concentric shells is filled with fiberglass or mineral wool. The circular silencer in Figure 5.4c has a center body and a double wall outer shell. Both rectangular and circular dissipative silencers come in several different pressure-drop configurations. The insertion loss, regenerated noise, and pressure drop of dissipative duct silencers are functions of silencer design and the location of the silencer in the duct system. These data are generally experimentally measured and are presented as part of manufacturers' data associated with their product lines. The data should be obtained in a manner consistent with the procedures outlined in ASTM Standard E477-84, "Standard Method of Testing Duct Liner Materials and Prefabricated Silencers for Acoustical and Airflow Performance" (53).

Active and dissipative duct silencers complement each other. Active silencers are usually effective between the 16 Hz and 250 Hz 1/1 octave frequency bands. Dissipative silencers are effective from the 63 Hz to 8000 Hz 1/1 octave frequency bands. The general insertion loss or attenuation characteristics of active and dissipative duct silencers are shown in Figure 5.5.

It is not practical to present data for a complete range of rectangular and circular duct silencers. These data are highly dependent on the manufacturer's design and will be different for each manufacturer. When possible, the silencer's manufacturer data should be used. If they are not available, the typical data presented for rectangular and circular, high and low pressure drop, dissipative silencers can be used to estimate the insertion loss, regenerated sound power, and pressure drop associated with selected rectangular and circular dissipative duct silencers. The

DUCT ELEMENT SOUND ATTENUATION

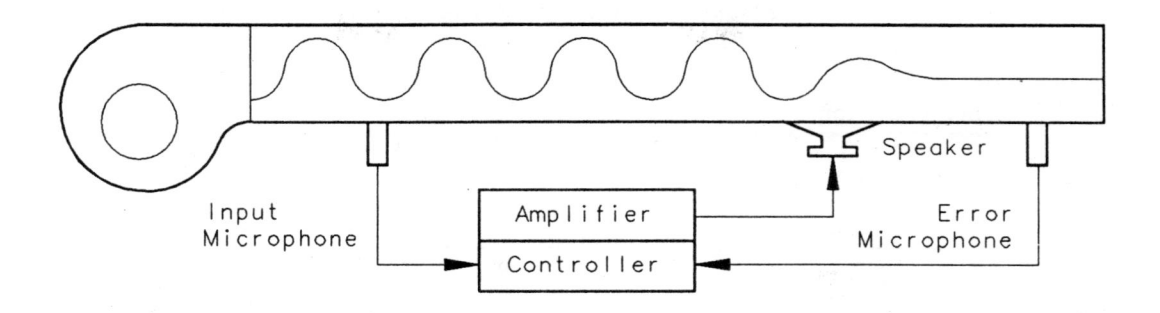

5–15

(a) Active Duct Silencer

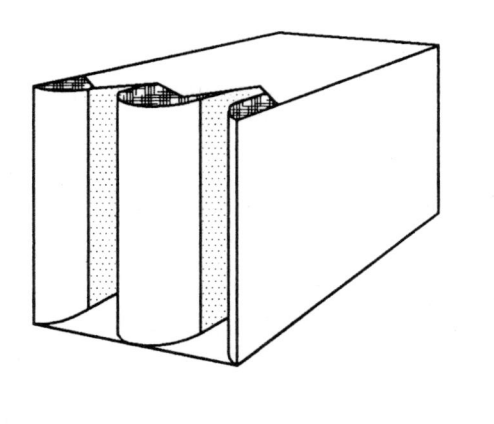

(b) Rectangular Dissipative Duct Silencer

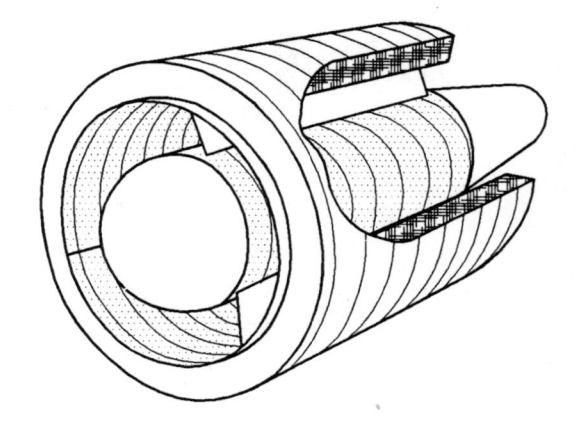

(c) Circular Dissipative Duct Silencer

Figure 5.4 Active and Dissipative Duct Silencers

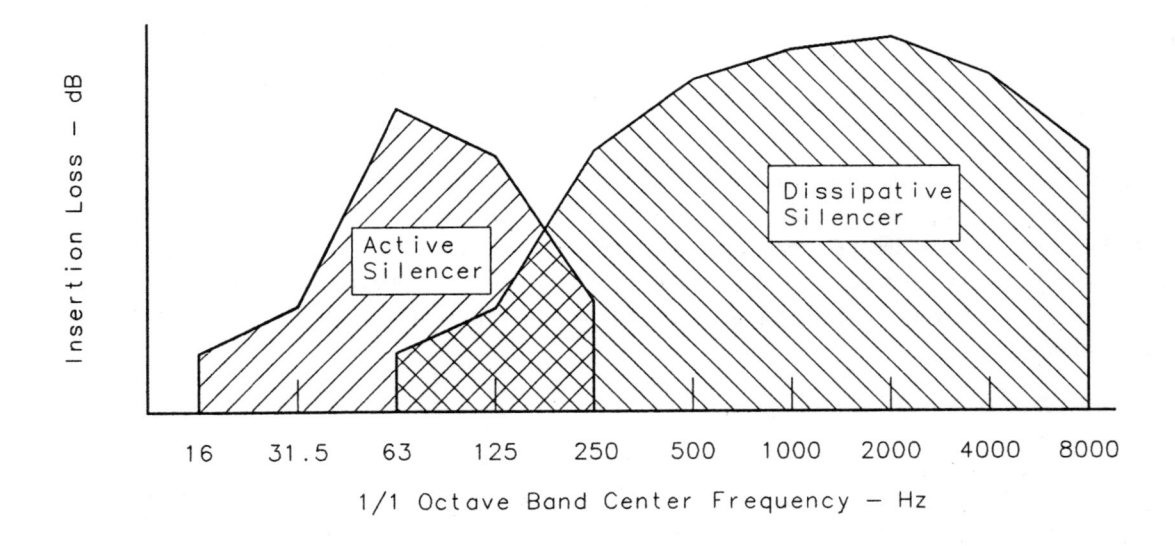

Figure 5.5 Insertion Loss of Active and Dissipative Duct Silencers

Table 5.16 7 ft, Rectangular, High–Pressure–Drop Duct Silencers

Face Vel. – fpm	63	125	250	500	1000	2000	4000	8000
	\multicolumn{8}{c}{Insertion Loss – dB}							
+1000	5	14	31	45	51	53	51	33
+2000	4	12	26	43	47	48	47	30
−1000	7	19	36	44	48	50	48	29
−2000	4	19	34	42	46	48	46	28
	\multicolumn{8}{c}{Regenerated Sound Power Levels – dB}							
+1000	58	52	42	36	37	35	30	30
+1500	64	57	58	49	45	49	48	46
+2000	72	65	64	63	55	56	57	56
−1000	55	52	54	55	55	64	64	55
−1500	61	58	59	61	61	66	75	66
−2000	69	66	65	75	71	73	79	76

Table 5.17 7 ft, Rectangular, Low–Pressure–Drop Duct Silencers

Face Vel. – fpm	63	125	250	500	1000	2000	4000	8000
	\multicolumn{8}{c}{Insertion Loss – dB}							
+1000	2	8	18	33	41	47	24	15
+2000	1	8	17	32	39	44	24	16
+2500	1	7	17	30	37	42	24	15
−1000	1	11	21	35	41	45	22	12
−2000	1	11	21	36	40	43	21	11
−2500	1	9	21	34	38	40	20	10
	\multicolumn{8}{c}{Regenerated Sound Power Levels – dB}							
+1000	58	51	40	34	35	28	27	19
+2000	67	61	58	53	51	54	52	45
+2500	74	68	65	62	56	59	60	55
−1000	58	49	46	44	49	45	34	25
−2000	70	61	59	56	57	62	58	50
−2500	74	67	64	62	61	65	65	57

data include insertion loss and regenerated sound power values for sound traveling with (+) and against (−) the airflow. Equations are presented that can be used to calculate the pressure loss across typical silencers and to calculate the silencer face area correction associated with regenerated sound power. Table 5.16 gives typical insertion loss and regenerated sound power levels for rectangular, high–pressure–drop duct silencers. Table 5.17 gives the same information for rectangular, low–pressure–drop silencers. The face area correction, FAC, for rectangular silencers is given by

$$FAC = 10 \log_{10} [FA] - 6 \tag{5.22}$$

where FA is the face area (ft^2) of the silencer. Table 5.18 gives typical insertion loss and regenerated sound power levels for circular, high–pressure–drop duct silencers. Table 5.19 gives the same information for circular, low–pressure–drop silencers. The face area correction, FAC, for circular duct silencers is given by

$$FAC = 10 \log_{10} [FA] - 4.76 \ . \tag{5.23}$$

DUCT ELEMENT SOUND ATTENUATION

Table 5.18 Circular, High–Pressure–Drop Duct Silencers (Double Wall with Center Body)

Face Vel. – fpm	63	125	250	500	1000	2000	4000	8000
1/1 Octave Band Center Frequency – Hz								
Insertion Loss – dB								
+1000	4	7	21	32	38	38	26	20
+2000	4	7	20	32	38	38	27	21
+3000	4	7	19	31	38	38	27	21
−1000	5	9	23	33	39	37	25	19
−2000	5	9	23	34	37	36	24	16
−3000	6	10	24	34	37	36	24	16
Regenerated Sound Power Levels – dB								
+1000	62	43	38	39	36	25	22	28
+2000	62	58	53	54	53	51	45	35
+3000	72	66	62	64	63	64	61	54
−1000	56	43	41	38	37	31	23	28
−2000	66	56	54	54	57	54	49	41
−3000	76	64	63	64	67	67	65	60

Table 5.19 Circular, Low Pressure Drop Duct Silencers (Double Wall with Center Body)

Face Vel. – fpm	63	125	250	500	1000	2000	4000	8000
1/1 Octave Band Center Frequency – Hz								
Insertion Loss – dB								
+1000	4	6	13	26	32	24	16	14
+2000	4	5	13	25	32	24	16	13
+3000	4	6	13	23	31	24	16	13
−1000	5	7	16	28	35	25	17	15
−2000	4	7	16	28	35	25	17	15
−3000	6	7	16	29	35	25	16	15
Regenerated Sound Power Levels – dB								
+1000	60	44	39	34	29	25	24	30
+2000	62	58	48	47	49	45	38	31
+3000	69	63	55	55	57	58	54	47
−1000	58	46	43	38	33	30	24	30
−2000	65	52	50	49	48	44	36	33
−3000	72	57	57	57	56	57	52	49

The silencer face velocity, V (fpm), is given by

$$V = Q/FA \tag{5.24}$$

where Q is the volume flow rate (cfm) and FA is as defined before.

The static pressure drop, ΔP (in. H_2O), across a rectangular duct silencer is obtained from

$$\Delta P = C_1 \cdot L^{C_2} \cdot C_3 \cdot V^{C_4} \tag{5.25}$$

where L is the silencer length (ft) and V is the silencer face velocity (fpm). The static pressure drop, ΔP, across a circular duct silencer is obtained from

$$\Delta P = C_3 \cdot \left[\frac{Q}{C_1 \cdot d^{C_2}} \right]^{C_4} \tag{5.26}$$

Table 5.20 Coefficients for Determining Static Pressure Drop across Duct Silencers

Rectangular	C_1	C_2	C_3	C_4
High Pressure Drop	0.6464	0.3971	2.637×10^{-7}	2.012
Low Pressure Drop	0.6015	0.4627	9.802×10^{-8}	2.011
Circular				
High Pressure Drop	$5.108 \cdot 10^{-3}$	1.999	$4.007 \cdot 10^{-8}$	2.002
Low Pressure Drop	$5.097 \cdot 10^{-3}$	2.000	$1.104 \cdot 10^{-8}$	2.022

Table 5.21 Coefficient for System Component Effect on Duct Silencers

Distance between silencer and closest edge of system component (fan, elbow, branch TO, etc.)	C(up) Upstream from entering edge of silencer	C(down) Downstream from leaving edge of silencer
$D_{eq} \times 3$ or greater	1.0	1.0
$D_{eq} \times 2$	1.4	1.4
$D_{eq} \times 1$	1.9	1.9
$D_{eq} \times 0.5$	3.0	3.0
Directly Connected	4.0	Not recommended

where Q is volume flow rate (cfm) through the silencer and d is the silencer face diameter (in.). The values of the coefficients C_1, C_2, C_3, and C_4 are given in Table 5.20. The pressure drops for dissipative duct silencers specified by Equations (5.25) and (5.26) are for the case where there are no system component effects associated with duct elements, such as fan discharge or return air sections, elbows, branch take-offs, etc., upstream or downstream of a duct silencer. When system component effects must be taken into account, a correction factor must be added to the pressure drops specified by Equations (5.25) and (5.26). The pressure drop, ΔP_s (in. H_2O), taking into account system component effects, is given by

$$\Delta P_s = \Delta P \cdot C_5 \tag{5.27}$$

where C_5 is obtained from the coefficients specified in Table 5.21 and is given by

$$C_5 = C(up) \times C(down) . \tag{5.28}$$

The equivalent duct diameter for circular ducts is the duct diameter. For rectangular ducts, the equivalent duct diameter, D_{eq} (in.) is

$$D_{eq} = \sqrt{4 \cdot W \cdot H / \pi} \tag{5.29}$$

where W is the width (in.) of the rectangular duct and H is the height (in.) of the rectangular duct.

When determining the effectiveness of a duct silencer, it is necessary to take into account both the insertion loss and the regenerated sound power levels of the silencer. If L_{W1} is the sound power level (dB) that exists before

DUCT ELEMENT SOUND ATTENUATION

the sound enters the silencer, the sound power level, L_{W2} (dB), at the exit of the silencer associated with the silencer insertion loss, IL (dB), is given by

$$L_{W2} = L_{W1} - IL . \tag{5.30}$$

The regenerated sound power levels, L_{W3}, associated with air flowing through a silencer are equal to the sound power levels, L_{Wr}, given in Tables 5.16, 5.17, 5.18, and 5.19 plus the face area correction, FAC, specified by Equation (5.22) or (5.23), or

$$L_{W3} = L_{Wr} + FAC . \tag{5.31}$$

The regenerated sound power level, L_{W3} (dB), must be added to L_{W2} to obtain the total sound power level, L_{W4} (dB), at the exit of the duct silencer. Because sound power levels are being added, L_{W2} and L_{W3} must be added logarithmically, or

$$L_{W4} = 10 \log_{10} \left[10^{(L_{W2}/10)} + 10^{(L_{W3}/10)} \right] . \tag{5.32}$$

EXAMPLE 5.9

A fan has the following sound power levels:

	\multicolumn{8}{c}{1/1 Octave Band Center Frequency — Hz}
	63	125	250	500	1000	2000	4000	8000
Fan L_{W1}, dB	91	87	83	82	78	76	72	70

The volume flow rate for the fan is 10,000 cfm and the fan has a total static pressure of 1.5 in. H_2O. If a low–pressure–drop rectangular duct silencer is used that has face dimensions of 30 in. by 24 in. and a length of 7 ft, determine the sound power level on the exit side of the duct silencer. There is a 90° elbow two duct diameters downstream from the elbow. The fan discharge section is located more than 3 duct diameters upstream from the silencer.

SOLUTION

Silencer face cross–sectional area $= \dfrac{30 \times 24}{144} = 5$ ft^2 .

$$V = \frac{10,000}{5} = 2,000 \ \frac{ft}{min} .$$

The static pressure drop from Equation 5.25 is

$$\Delta P = C_1 \cdot L^{C_2} \cdot C_3 \cdot V^{C_4}$$

where $C_1 = 0.6015$, $C_2 = 0.4627$, $C_3 = 9.802 \cdot 10^{-8}$, and $C_4 = 2.011$. Thus,

$$\Delta P = 0.6016 \cdot (7)^{0.4627} \cdot 9.802 \cdot 10^{-8} \cdot (2,000)^{2.011} = 0.63 \ in. \ H_2O .$$

From Table 5.21, C(up) equals 1 and C(down) equals 1.4. Thus, C_5 equals 1.4 and

$$\Delta P_s = 1.4 \times 0.63 = 0.88 \ in. \ H_2O .$$

The face area adjustment factor Equation (5.22) is

$$FAC = 10 \log_{10} [5] - 6 = 1.0 \ dB .$$

The results are tabulated below.

	1/1 Octave Band Center Frequency — Hz							
	63	125	250	500	1000	2000	4000	8000
Fan L_{W1}, dB	91	87	83	82	78	76	72	70
Silencer IL, dB	−4	−12	−26	−43	−47	−48	−47	−30
L_{W2}, dB	87	75	57	39	31	28	25	30
Flow L_{Wr}, dB	72	65	64	63	55	56	57	56
FAC, dB	1	1	1	1	1	1	1	1
L_{W3}, dB	73	66	65	64	56	57	58	57
L_{W4}, dB	87	76	66	64	56	57	58	57

5.9 DUCT BRANCH SOUND POWER DIVISION

When sound traveling in a duct encounters a junction, the sound power contained in the incident sound waves in the main feeder duct is distributed between the branches associated with the junction. This division of sound power is referred to as the branch sound power division. The corresponding attenuation of sound power that is transmitted down each branch of the junction has of two components (44,46). The first is associated with the reflection of the incident sound wave if the sum of the cross-sectional areas of the individual branches, ΣS_{Bi}, differs from the cross-sectional area, S_M, of the main feeder duct. The second component is associated with the ratio of the cross-sectional area, S_{Bi}, of an individual branch divided by the sum of the cross-sectional areas of the individual branches, ΣS_{Bi}. The attenuation of sound power, ΔL_{Bi}, at a junction that is related to the sound power transmitted down an individual branch of the junction is given by

$$\Delta L_{B_i} = 10 \log_{10} \left[1 - \left[\frac{\frac{\Sigma S_{Bi}}{S_M} - 1}{\frac{\Sigma S_{Bi}}{S_M} + 1} \right]^2 \right] + 10 \log_{10} \left[\frac{S_{Bi}}{\Sigma S_{Bi}} \right] \quad (5.33)$$

where S_{Bi} is the cross-sectional area of branch i, ΣS_{Bi} is the total cross-sectional area of the individual branches that continue from the main feeder duct, and S_M is the cross-sectional area of the main duct. The first term in Equation (5.33) is related to the reflection of the incident wave when the area of the branches differs from the area of the main feeder duct. It is present only when the sound waves propagating in the main feeder duct are plane waves and when ΣS_{Bi} is not equal to S_M. Plane wave propagation in a duct exists at frequencies below

$$f_{co} = \frac{c_o}{2a} \quad (5.34)$$

where c_o is the speed of sound in air (ft/s) and a is the larger cross-sectional dimension (ft) of a rectangular duct, or

$$f_{co} = 0.586 \frac{c_o}{d} \quad (5.35)$$

where d is the diameter (ft) of a circular duct (27). The cutoff frequency, f_{co}, is the frequency above which plane waves no longer propagate in a duct. At these higher frequencies, the waves that propagate in the duct are referred to as cross or spinning modes. The second term in Equation (5.33) is associated with the division of the remaining incident sound power at the junction between the individual branches. If the total cross-sectional area of the branches after the junction is equal to the cross-sectional area of the main feeder duct or if the frequencies of interest are above the cutoff frequency, Equation (5.33) reduces to

$$\Delta L_{B_i} = 10 \log_{10}\left[\frac{S_{Bi}}{\Sigma S_{Bi}}\right]. \tag{5.36}$$

EXAMPLE 5.10

An 18-in.-diameter main feeder duct terminates into a junction that has a 12-in.-diameter branch (continuation of the main feeder duct) and a 6-in.-diameter 90° branch takeoff. Determine the attenuation (dB) of the sound power transmitted into the 90° branch takeoff.

SOLUTION

$S_M = \frac{\pi \cdot 18^2}{4} = 254.5 \text{ in.}^2$.   Main feeder duct

$S_{B1} = \frac{\pi \cdot 12^2}{4} = 113.1 \text{ in.}^2$.   Continuation branch

$S_{B2} = \frac{\pi \cdot 6^2}{4} = 28.3 \text{ in.}^2$.   90° branch takeoff

$\Sigma S_{Bi} = 113.1 + 28.3 = 141.4 \text{ in.}^2$.

$f_{co} = \frac{0.586 \cdot 1125 \cdot 12}{18} = 439.5 \text{ Hz}$.

Using Equation 5.33, the branch power division associated with branch 2 can be determined:

$\frac{S_{B2}}{\Sigma S_{Bi}} = \frac{28.3}{141.4} = 0.2$.

$1 - \left[\frac{\frac{\Sigma S_{Bi}}{S_M} - 1}{\frac{\Sigma S_{Bi}}{S_M} + 1}\right]^2 = 1 - \left[\frac{\frac{141.4}{254.5} - 1}{\frac{141.4}{254.5} + 1}\right]^2 = 0.918$.

The results are tabulated below.

	63	125	250	500	1000	2000	4000	8000
$10 \log_{10}[0.918]$, dB	0.4	0.4	0.4					
$10 \log_{10}[0.2]$, dB	7.0	7.0	7.0	7.0	7.0	7.0	7.0	7.0
ΔL_{B2}, dB	7.4	7.4	7.4	7.0	7.0	7.0	7.0	7.0

(1/1 Octave Band Center Frequency — Hz)

5.10 DUCT-END REFLECTION LOSS

When low-frequency plane sound waves interact with a small diffuser that discharges into a large room, a significant amount of the sound energy incident on this interface is reflected back into the duct. This interaction results in substantial low-frequency sound attenuation. Very few experimental data, relative to the low-frequency sound attenuation associated with duct end reflection losses, are available in the open literature. One of the more significant works is that published by Sandbakken et al. (32). This work resulted in the technical basis for the duct-end reflection correction that is part of AMCA Standard 300-85, "Reverberation Room Method for Sound Testing of Fans" (1). The report describing the above work presents sound attenuation information that is associated with ducts terminated in free space (ducts terminated more than three duct diameters away from a reflecting surface) and ducts terminated flush with a wall. The sound attenuation, ΔL, associated with duct–end reflection losses presented by the above authors and in AMCA Standard 300–85 can be approximated by

$$\Delta L = 10 \log_{10} \left[1 + \left[\frac{c_o}{\pi \cdot f \cdot D} \right]^{1.88} \right] \quad (5.37)$$

for ducts terminated in free space and by

$$\Delta L = 10 \log_{10} \left[1 + \left[\frac{0.8 \cdot c_o}{\pi \cdot f \cdot D} \right]^{1.88} \right] \quad (5.38)$$

for ducts terminated flush with a wall, where f is frequency (Hz), c_o is the speed of sound in air (ft/s), and D is the diameter (ft) of a circular duct or the effective diameter of a rectangular duct. If the duct is rectangular, D is

$$D = \left[\frac{4 \cdot \text{Area}}{\pi} \right]^{1/2} \quad (5.39)$$

where Area is the area (ft^2) of the rectangular duct. D can have the unit of inches if c_o has the units of in/s.

There are some limitations associated with Equations (5.37) and (5.38). The tests reported by Sandbakken et al. were for straight sections of circular ducts. These ducts directly terminated into a reverberation chamber with no restriction on the end of the duct or with a circular orifice constriction placed over the end of the duct. Diffusers can be either round or rectangular. They usually always have a restriction associated with them thath may be either a damper, guide vanes to direct airflow, a perforated metal facing, or a combination of these elements. Currently, there are no data that indicate the effects of these elements. It is not known whether these elements react similar to the orifices used in the above–described project. As a result, the effects of an orifice placed over the end of a duct are not included in Equations (5.37) and (5.38). One can assume that using Equation (5.39) to calculate D will yield reasonable results with diffusers that have low aspect ratios (length/width). However, many types of diffusers (particularly slot diffusers) have high aspect ratios. It is currently not known whether Equations (5.37) or (5.38) can be accurately used with these diffusers. Finally, many diffusers do not have long straight sections (greater than three duct diameters) before they terminate into a room. Many duct sections between a main feed branch and a diffuser may be curved or may be short, stubby sections. The effects of these configurations on the duct–end reflection loss are currently not known. It is felt that Equations (5.37) and (5.38) can be used with reasonable accuracy for many diffuser configurations. However, some caution should be exercised when a diffuser configuration differs drastically from the test conditions used to derive these equations.

EXAMPLE 5.11

Determine the duct–end reflection loss associated with a circular diffuser that has a diameter of 12 in. Assume the diffuser terminates in free space.

SOLUTION

Use Equation (5.37) for the calculations associated with this example. The results are tabulated below.

	\multicolumn{8}{c}{1/1 Octave Band Center Frequency – Hz}
	63	125	250	500	1000	2000	4000	8000
$c_o/(\pi \cdot f \cdot D)$	5.72	2.86	1.43	0.72	0.36	0.18	0.09	0.04
ΔL, dB	14.4	9.1	4.7	1.9	0.6	0.2	0.0	0.0

5.11 TERMINAL VOLUME REGULATION UNITS

Currently there are no data in the open literature relative to the sound attenuation associated with terminal volume regulation units. Most of the available data will be proprietary manufacturers' product data. The insertion loss values listed below are from the sound and vibration control chapter of the ASHRAE 1984 Systems Handbook (38).

Table 5.21 Typical Insertion Loss Values of Terminal Volume Regulation Units

	\multicolumn{8}{c}{1/1 Octave Band Center Frequency — Hz}
	63	125	250	500	1000	2000	4000	8000
Insertion Loss, dB	0	5	10	15	15	15	15	15

DUCT BREAKOUT AND BREAKIN — CHAPTER 6

6.1 SOUND BREAKOUT AND BREAKIN

Noise that is generated within a duct and then transmitted through the duct wall into the surrounding area is called "breakout" (Figure 6.1a). This phenomenon is often referred to as low–frequency duct rumble. There are two possible sources for duct breakout. One is associated with noise that is generated within the duct, usually by a fan. This noise, designated W_i in Figure 6.1a, is transmitted down the duct and then through the duct walls into surrounding spaces. The transmitted sound is designated W_r in Figure 6.1a. The second source is associated with turbulent airflow that aerodynamically excites the duct walls, causing them to vibrate. This vibration generates low–frequency duct rumble, which is then radiated into the surrounding spaces. In many situations, particularly near fan discharge sections, duct breakout may be associated with both of these sources. The information on duct breakout that is presented in this chapter is based on work reported by Cummings (11,12,13,36,38). This work dealt only with sound that was generated within the duct (by loud speakers) and then transmitted through the duct walls. Cummings' work did not cover aerodynamically excited duct rumble.

Noise that is transmitted into a duct from the surrounding area and then transmitted within the duct is called "breakin" (Figure 6.1b). W_i in the figure refers to sound in the area surrounding a duct that is incident on the duct walls; W_t refers to the sound that is transmitted within the duct.

The breakout transmission loss, TL_{out} (dB), of a duct is given by

$$TL_{out} = 10 \log_{10} \left[\frac{W_i}{A_i} \frac{A_o}{W_r} \right] \qquad (6.1)$$

where W_i is the sound power (watts) in the duct, W_r is the sound power (watts) radiated from the duct, A_i is the cross–sectional area (in.^{2}) of the inside of the duct, and A_o is the sound radiation surface area (in.^{2}) of the outside of the duct (11,12,19,36). Rearranging Equation (6.1) yields

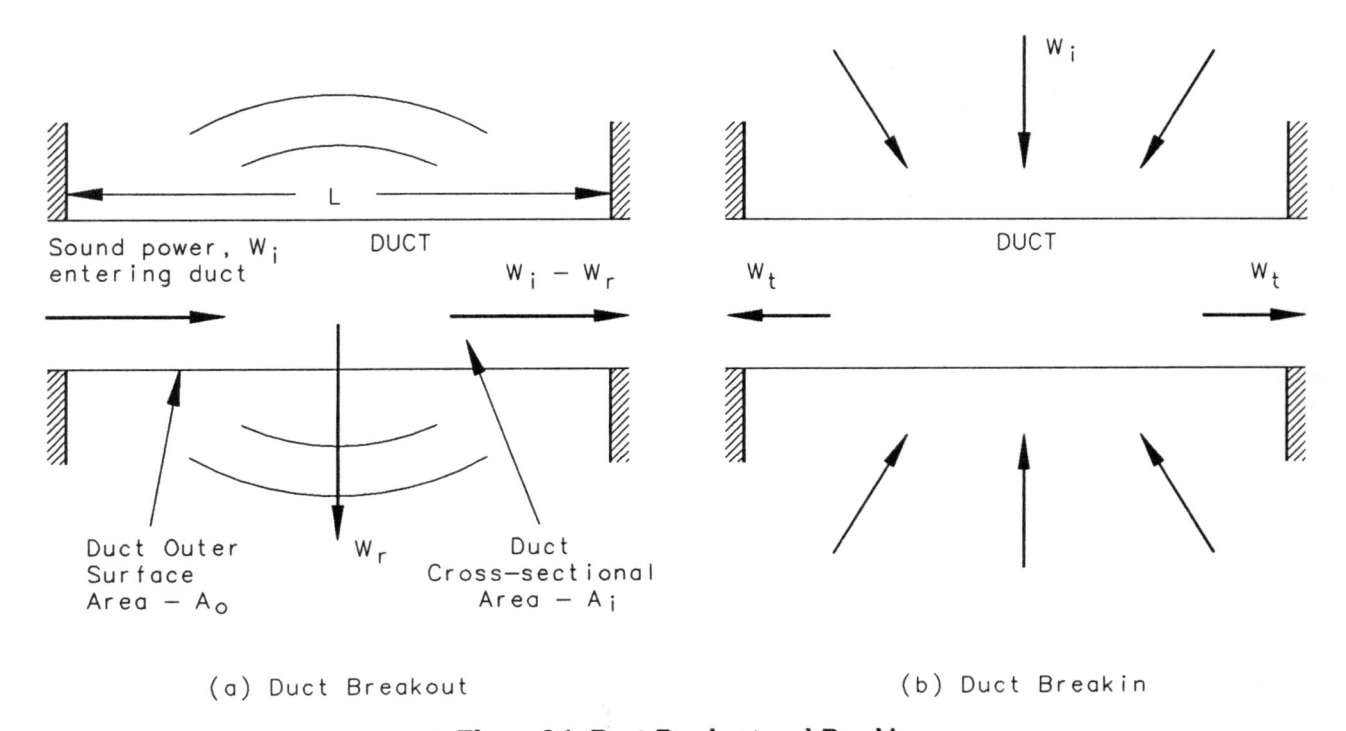

(a) Duct Breakout   (b) Duct Breakin

Figure 6.1 Duct Breakout and Breakin

$$L_{W_r} = L_{W_i} + 10 \log_{10} \left[\frac{A_o}{A_i}\right] - TL_{out} \qquad (6.2)$$

where L_{W_r} (dB) and L_{W_i} (dB) are given by

$$L_{W_r} = 10 \log_{10} \left[\frac{W_r}{10^{-12}}\right] \qquad (6.3)$$

$$L_{W_i} = 10 \log_{10} \left[\frac{W_i}{10^{-12}}\right]. \qquad (6.4)$$

The breakin transmission loss, TL_{in} (dB), associated with ducts is given by

$$TL_{in} = 10 \log_{10} \left[\frac{W_i}{2 \cdot W_t}\right] \text{ dB} \qquad (6.5)$$

where W_i is the incident sound power (watts) on the duct from the surrounding space and W_t is the sound power (watts) that travels along the duct both upstream and downstream from the point where the sound enters the duct. The sound power level of the sound transmitted into the duct is obtained by rearranging Equation (6.5), or

$$L_{W_t} = L_{W_i} - TL_{in} - 3 \text{ dB} \qquad (6.6)$$

where L_{W_i} is given by Equation (6.4) and L_{W_t} is given by

$$L_{W_t} = 10 \log_{10} \left[\frac{W_t}{10^{-12}}\right] \text{ dB}. \qquad (6.7)$$

6.2 RECTANGULAR DUCTS

If the duct is a rectangular duct, A_i and A_o in Equations (6.1) and (6.2) are given by

$$A_i = a \cdot b \qquad (6.8)$$

$$A_o = 24 \cdot L \cdot (a + b) \qquad (6.9)$$

where a is the larger duct cross-sectional dimension (in.), b is the smaller duct cross-sectional dimension (in), and L is the exposed length (ft) of the duct (Figure 6.1a). For rectangular ducts, the breakout transmission loss curve shown in Figure 6.2 can be divided into two regions: (1) a region where plane mode transmission within the duct is dominant and (2) a region where multi-mode transmission is dominant. The frequency, f_L, that divides these two regions is given by

$$f_L = \frac{24134}{\sqrt{a \cdot b}}. \qquad (6.10)$$

If $f < f_L$, the plane mode predominates and TL_{out} is given by

$$TL_{out} = 10 \log_{10} \left[\frac{f \cdot q^2}{a + b}\right] + 17 \qquad (6.11)$$

where f is frequency (Hz), q is the mass/unit area (lb/ft^2) of the duct walls, and a and b are as described above. If $f \geq f_L$, multi-mode transmission predominates and TL_{out} is calculated from

DUCT BREAKOUT AND BREAKIN

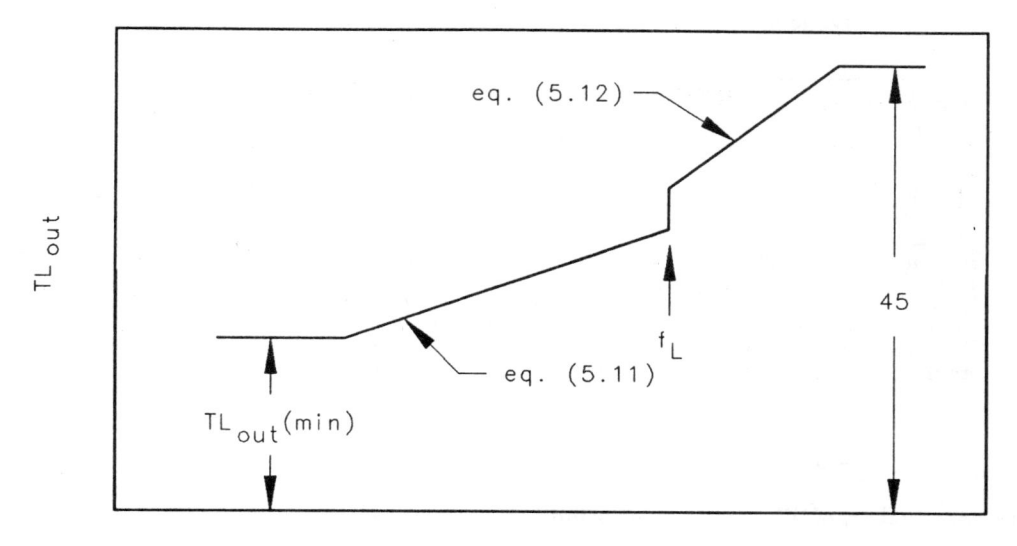

6–3

Figure 6.2 TL_{out} Associated with Rectangular Ducts

Table 6.1 TL_{out} vs. Frequency for Various Rectangular Ducts

Duct Size (in.)	Gauge	63	125	250	500	1000	2000	4000	8000
12 x 12	24 ga.	21	24	27	30	33	36	41	45
12 x 24	24 ga.	19	22	25	28	31	35	41	45
12 x 48	22 ga.	19	22	25	28	31	37	43	45
24 x 24	22 ga.	20	23	26	29	32	37	43	45
24 x 48	20 ga.	20	23	26	29	31	39	45	45
48 x 48	18 ga.	21	24	27	30	35	41	45	45
48 x 96	18 ga.	19	22	25	29	35	41	45	45

(1/1 Octave Band Center Frequency — Hz)

Data are for duct lengths of 20 ft, but values may be used for the cross section shown regardless of length.

$$TL_{out} = 20 \cdot \log_{10}[q \cdot f] - 31 \text{ dB} \tag{6.12}$$

where q and f are as specified above. The minimum value of TL_{out} occurs when $W_i = W_r$ and is specified by

$$TL_{out}(min) = 10 \log_{10}\left[24 \cdot L \cdot \left[\frac{1}{a} + \frac{1}{b}\right]\right]. \tag{6.13}$$

Table 6.1 shows some values of TL_{out} calculated using the above equations.

The breakin transmission loss can be divided into two regions, which are separated by a cutoff frequency, f_1. The cutoff frequency is the frequency for the lowest acoustic cross–mode in the duct. It is given by

$$f_1 = \frac{6764}{a}. \tag{6.14}$$

If $f \leq f_1$,

$$TL_{in} = \text{larger of} \begin{cases} TL_{out} - 4 - 10 \log_{10}\left[\frac{a}{b}\right] + 20 \log_{10}\left[\frac{f}{f_1}\right] & (6.15a) \\ 10 \log_{10}\left[12 \cdot L \cdot \left[\frac{1}{a} + \frac{1}{b}\right]\right]. & (6.15b) \end{cases}$$

Table 6.2 TL_{in} vs. Frequency for Various Rectangular Ducts

Duct Size (in.)	Gauge	63	125	250	500	1000	2000	4000	8000
12 x 12	24 ga.	16	16	16	25	30	33	38	42
12 x 24	24 ga.	15	15	17	25	28	32	38	42
12 x 48	22 ga.	14	14	22	25	28	34	40	42
24 x 24	22 ga.	13	13	21	26	29	34	40	42
24 x 48	20 ga.	12	15	23	26	28	36	42	42
48 x 48	18 ga.	10	19	24	27	32	38	42	42
48 x 96	18 ga.	11	19	22	26	32	38	42	42

1/1 Octave Band Center Frequency — Hz

Data are for duct lengths of 20 ft, but values may be used for the cross section shown regardless of length.

If $f > f_1$,
$$TL_{in} = TL_{out} - 3 \text{ dB} . \tag{6.16}$$

Table 6.2 shows some values of TL_{in} calculated using the above equations.

EXAMPLE 6.1

Determine the breakout and breakin sound power for a duct with the following dimensions: smaller duct dimension — 12 in.; larger duct dimension — 24 in.; duct length — 20 ft. The duct is constructed of 24 gauge sheet metal.

SOLUTION

q = mass/unit area of 24 ga sheet metal = 1.0 lb/ft^2.

Breakout:

$A_i = a \cdot b = 24 \cdot 12 = 288 \text{ in}^2$ $\qquad A_o = 24 \cdot 20 \cdot (24 + 12) = 17,280 \text{ in}^2$.

$$f_L = \frac{24134}{\sqrt{24 \cdot 12}} = 1,422 \text{ Hz} .$$

$$TL_{out}(min) = 10 \log_{10} \left[24 \cdot 20 \cdot \left[\tfrac{1}{24} + \tfrac{1}{12} \right] \right] = 17.8 \text{ dB} .$$

$10 \log_{10} \left[A_o / A_i \right] = 10 \log_{10} [17,280/288] = 17.8 .$

The results are tabulated below.

	63	125	250	500	1000	2000	4000	8000
Eq. 6.11	19.4	22.4	25.4	28.4	31.4			
Eq. 6.12						35.0	41.0	45.0
TL_{out}, dB	19.4	22.4	25.4	28.4	31.4	35.0	41.0	45.0
$10 \log_{10} [A_o/A_i]$	17.8	17.8	17.8	17.8	17.8	17.8	17.8	17.8
$L_{W_r} - L_{W_i}$, dB	−1.6	−4.6	−7.6	−10.6	−13.6	−17.2	−23.2	−27.2

1/1 Octave Band Center Frequency — Hz

Breakin:

$f_1 = 6764/24 = 282 \text{ Hz} .$

The results are tabulated below.

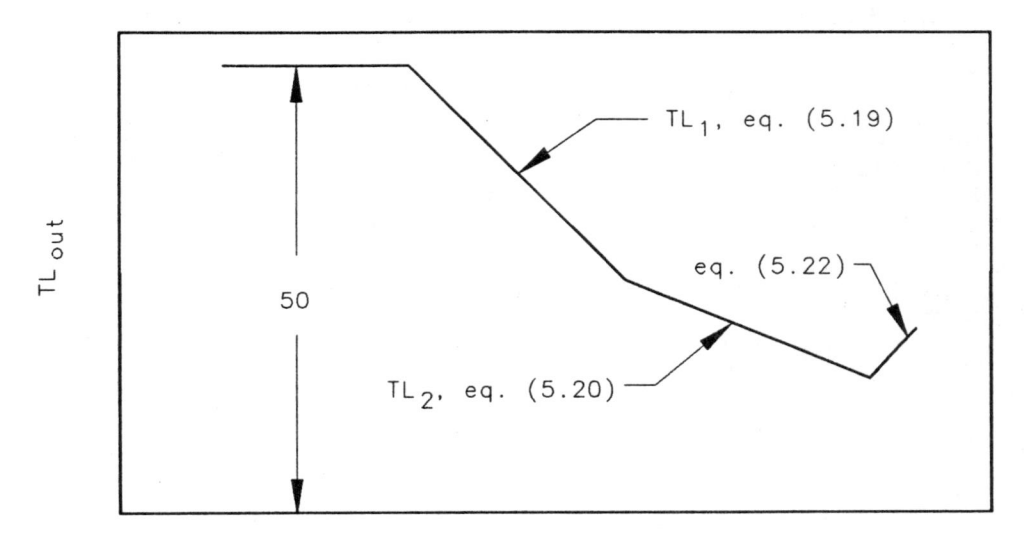

DUCT BREAKOUT AND BREAKIN

Figure 6.3 TL_{out} Associated with Circular Ducts

	\multicolumn{8}{c}{1/1 Octave Band Center Frequency – Hz}
	63	125	250	500	1000	2000	4000	8000
TL_{out}, dB	19.2	22.4	25.4	28.4	31.4	35.0	41.0	45.0
Eq. 6.15a	−1.0	8.3	17.3					
Eq. 6.15b	14.8	14.8	14.8					
	14.8	14.8	17.3					
				−3.0	−3.0	−3.0	−3.0	−3.0
TL_{in}, dB	14.8	14.8	17.3	25.4	28.4	32.0	38.0	42.0
	3.0	3.0	3.0	3.0	3.0	3.0	3.0	3.0
$L_{W_t} - L_{W_i}$, dB	−17.8	−17.8	−20	−28.4	−31.4	−35	−41	−45

6.3 CIRCULAR DUCTS

If the duct is circular, A_i and A_o in Equations (6.1) and (6.2) are given by

$$A_i = \pi \frac{d^2}{4} \tag{6.17}$$

$$A_o = 12 \cdot L \cdot \pi \cdot d \tag{6.18}$$

where d is the duct diameter (in.) and L is the exposed length (ft) of the duct. Narrow band and 1/3 octave band breakout transmission loss values for circular ducts are very hard to predict, and no simple prediction techniques are available (11,12). However, if the analysis is limited to 1/1 octave frequency bands, TL_{out} associated with circular ducts can be approximated by a curve similar to the one shown in Figure 6.3. Table 6.3 shows TL_{out} data for circular ducts that were obtained by Cummings (11,12). If the breakout analysis is limited to 1/1 octave band values, Equations (6.19) and (6.20) can be used to approximate the data in Table 6.3.

Table 6.3 Experimentally Measured TL_{out} vs. Frequency for Circular Ducts

Diam.	Length, gauge	63	125	250	500	1000	2000	4000
\multicolumn{9}{l}{Long Seam Ducts}								
8 in.	15 ft, 26 ga.	45	53	55	52	44	35	34
14 in.	15 ft, 24 ga.	50	60	54	36	34	31	25
22 in.	15 ft, 22 ga.	47	53	37	33	33	27	25
32 in.	15 ft, 22 ga.	51	46	26	26	24	22	38
\multicolumn{9}{l}{Spiral Wound Ducts}								
8 in.	10 ft, 26 ga.	48	64	75	72	56	56	46
14 in.	10 ft, 26 ga.	43	53	55	33	34	35	25
26 in.	10 ft, 24 ga.	45	50	26	26	25	22	36
26 in.	10 ft, 16 ga.	48	53	36	32	32	28	41
32 in.	10 ft, 22 ga.	43	42	28	25	26	24	40

1/1 Octave Band Center Frequency — Hz

Table 6.4 Calculated TL_{out} vs. Frequency for Circular Ducts

Diam.	Length, gauge	63	125	250	500	1000	2000	4000
\multicolumn{9}{l}{Long Seam Ducts}								
8 in.	15 ft, 26 ga.	50	50	50	44	42	40	38
14 in.	15 ft, 24 ga.	50	50	48	37	35	33	31
22 in.	15 ft, 22 ga.	50	50	38	32	30	28	26
32 in.	15 ft, 22 ga.	50	44	29	26	24	22	37
\multicolumn{9}{l}{Spiral Wound Ducts}								
8 in.	10 ft, 26 ga.	50	50	50	46	42	40	38
14 in.	10 ft, 26 ga.	50	50	48	35	33	31	29
26 in.	10 ft, 24 ga.	45	45	35	27	25	23	38
26 in.	10 ft, 16 ga.	50	50	42	34	32	30	45
32 in.	10 ft, 22 ga.	50	47	32	26	24	22	37

$$TL_1 = 17.6 \log_{10}[q] - 49.8 \log_{10}[f] - 55.3 \log_{10}[d] + C_o \qquad (6.19)$$

$$TL_2 = 17.6 \log_{10}[q] - 6.6 \log_{10}[f] - 36.9 \log_{10}[d] + 97.4 \qquad (6.20)$$

where q is the mass/unit area (lb/ft^2) of the duct wall, f is frequency (Hz), d is the inside duct diameter (in.), and

$C_o = 230.4$ for long seam ducts

$C_o = 232.9$ for spiral wound ducts

TL_{out} = the larger of $TL_{1,2}$. $\qquad (6.21)$

The above equations yield good results except when the diameter of the duct is greater than 26 inches and the 1/1 octave band center frequency is equal to 4000 Hz. For this special case, TL_{out} is given by

$$TL_{out} = 17.6 \log_{10}[q] - 36.9 \log_{10}[d] + 90.6 . \qquad (6.22)$$

The maximum allowable value for TL_{out} is 50 dB. Thus, if the value for TL_1 obtained from Equation (6.19) exceeds 50 dB, the value should be set equal to 50 dB. Table 6.4 lists the calculated values for TL_{out} that correspond to the values listed in Table 6.3.

For calculating the breakin transmission loss for circular ducts, the cut–off frequency for the lowest acoustic cross–mode is given by

DUCT BREAKOUT AND BREAKIN

Table 6.5 Experimentally Determined TL_{in} vs. Frequency for Circular Ducts

Diam.	Length, gauge	63	125	250	500	1000	2000	4000
Long Seam Ducts								
8 in.	15 ft, 26 ga.	17	31	39	42	41	32	31
14 in.	15 ft, 24 ga.	27	43	43	31	31	28	22
22 in.	15 ft, 22 ga.	28	40	30	30	30	24	22
32 in.	15 ft, 22 ga.	35	36	23	23	21	19	35
Spiral Wound Ducts								
8 in.	10 ft, 26 ga.	20	42	59	62	53	43	26
14 in.	10 ft, 26 ga.	20	36	44	28	31	32	22
26 in.	10 ft, 24 ga.	27	38	20	23	22	19	33
26 in.	10 ft, 16 ga.	30	41	30	29	29	25	38
32 in.	10 ft, 22 ga.	27	32	25	22	23	21	37

Table 6.6 Calculated TL_{in} vs. Frequency for Circular Ducts

Diam.	Length, gauge	63	125	250	500	1000	2000	4000
Long Seam Ducts								
8 in.	15 ft, 26 ga.	17	23	29	34	39	37	35
14 in.	15 ft, 24 ga.	22	28	34	32	32	30	28
22 in.	15 ft, 22 ga.	26	32	31	29	27	25	23
32 in.	15 ft, 22 ga.	29	34	26	23	21	19	34
Spiral Wound Ducts								
8 in.	10 ft, 26 ga.	17	23	29	35	39	37	35
14 in.	10 ft, 26 ga.	27	38	37	30	30	28	26
26 in.	10 ft, 24 ga.	27	33	29	24	22	20	35
26 in.	10 ft, 16 ga.	27	33	36	31	29	27	42
32 in.	10 ft, 22 ga.	29	35	29	23	21	19	34

$$f_1 = \frac{7929}{d} \quad . \tag{6.23}$$

If $f \leq f_1$,

$$TL_{in} = \text{larger of} \begin{cases} TL_{out} - 4 + 20\,Log_{10}\left[\dfrac{f}{f_1}\right] & \text{(6.24a)} \\ 10\,Log_{10}\left[\dfrac{2L}{d}\right] \quad . & \text{(6.24b)} \end{cases}$$

If $f > f_1$, the breakin transmission loss is defined by

$$TL_{in} = TL_{out} - 3 \text{ dB} \quad . \tag{6.25}$$

Table 6.5 gives experimentally determined values for the breakin transmission loss for various duct sizes, and Table 6.6 gives the corresponding values calculated using the above equations.

EXAMPLE 6.2

Determine the breakout and breakin sound power of a long seam circular duct given the following information: diameter — 14 in.; length — 15 ft. The duct is constructed of 24 gauge sheet metal.

SOLUTION

q = mass/unit area of 24 ga sheet metal = 1.0 lb_m/ft^2.

Breakout:

$$A_i = \pi \frac{14^2}{4} = 153.9 \text{ in}^2 \qquad A_o = 12 \cdot 15 \cdot \pi \cdot 14 = 7{,}916.8 \text{ in}^2.$$

$$10 \log_{10}[A_o/A_i] = 10 \log_{10}(7{,}916.8/153.9) = 17.1 \text{ dB}.$$

The results are tabulated below.

	\multicolumn{7}{c}{1/1 Octave Band Center Frequency — Hz}
	63	125	250	500	1000	2000	4000
TL_1, dB	77.6	62.6	47.6	32.6	17.6	2.6	−12.4
TL_2, dB	43.3	41.3	39.3	37.3	36.3	33.3	31.3
TL_{max}, dB	50.0	50.0					
TL_{out}, dB	50.0	50.0	47.6	37.3	35.3	33.3	31.3
$10 \log_{10}[A_o/A_i]$	17.1	17.1	17.1	17.1	17.1	17.1	17.1
$L_{W_r} - L_{W_i}$, dB	−32.9	−32.9	−30.5	−20.2	−18.2	−16.2	−14.2

Breakin:

$$f_1 = 7929/14 = 566.4 \text{ Hz}.$$

The results are tabulated below.

	\multicolumn{7}{c}{1/1 Octave Band Center Frequency — Hz}
	63	125	250	500	1000	2000	4000
TL_{out}, dB	45.0	45.0	45.6	37.3	35.3	33.3	31.3
Eq. 6.24a	26.9	32.9	36.5	32.2			
Eq. 6.24b	14.1	14.1	14.1	14.1			
	26.9	32.9	36.5	32.2			
					−3.0	−3.0	−3.0
TL_{in}, dB	26.9	32.9	36.5	32.2	32.3	30.3	28.3
	3.0	3.0	3.0	3.0	3.0	3.0	3.0
$L_{W_t} - L_W$, dB	−29.9	35.9	−39.5	−35.2	−35.3	−33.3	−31.3

6.4 FLAT OVAL DUCTS

If the duct is a flat oval duct, A_i and A_o in equations (6.1) and (6.2) are given by

$$A_i = b \cdot (a - b) + \frac{\pi b^2}{4} \tag{6.26}$$

$$A_o = 12 \cdot L \cdot [2(a - b) + \pi b] \tag{6.27}$$

$$P = 2 \cdot (a - b) + \pi b \tag{6.28}$$

where a is the length (in.) of the major duct axis, b is the length (in.) of the minor duct axis, L is the duct length

DUCT BREAKOUT AND BREAKIN

(ft), A_i is the cross-sectional area (in.²), A_o is the surface area of the outside of the duct (in.²), and P is the perimeter of the duct in inches (Figure 6.4). The fraction of the perimeter taken up by the the flat sides, σ, is given by

$$\sigma = 1/\left[1 + \frac{\pi b}{2(a-b)}\right] . \tag{6.29}$$

The minimum breakout transmission loss, $TL_{out}(min)$ (dB), for flat oval ducts is given by

$$TL_{out}(min) = 10 \log_{10}\left[\frac{A_o}{A_i}\right] . \tag{6.30}$$

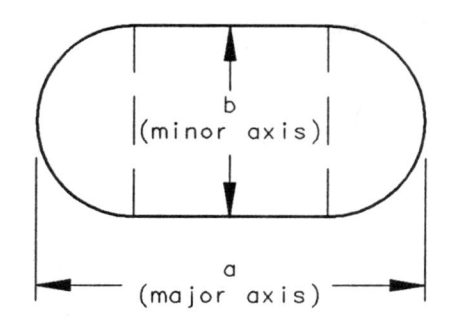

Figure 6.4 Flat Oval Duct

The low-to-mid-frequency transmission loss, TL_{out} (dB), associated with flat oval ducts is specified by

$$TL_{out} = 10 \cdot \log_{10}\left[\frac{q^2 f}{\sigma^2 P}\right] + 20 \text{ dB} . \tag{6.31}$$

The upper frequency limit, f_L (Hz), of applicability of Equation (6.31) is

$$f_L = \frac{8115}{b} . \tag{6.32}$$

Table 6.7 gives some values of TL_{out} for flat oval ducts of various sizes.

As was the case with rectangular and circular ducts, TL_{in} can be written in terms of TL_{out}. While there are no exact solutions for the cut-off frequency for the lowest acoustic cross-mode in flat oval ducts, equation (6.33) gives an approximate solution:

$$f_1 = \frac{6764}{(a-b)\cdot\left[1 + \frac{\pi b}{2(a-b)}\right]^{1/2}} \tag{6.33}$$

where a and b are in inches. This equation is valid when $a/b \geq 2$. When $a/b < 2$, the accuracy of Equation (6.43) deteriorates progressively as a/b approaches unity. When $f \leq f_1$, TL_{in} is given by

$$TL_{in} = \text{the larger of} \begin{cases} TL_{out} + 10 \text{ Log}_{10}\left[f^2 \cdot A_i\right] - 81 & (6.34a) \\ 10 \text{ Log}_{10}\left[\frac{6 \cdot P \cdot L}{A_i}\right] . & (6.34b) \end{cases}$$

When $f > f_1$, TL_{in} is given by

$$TL_{in} = TL_{out} - 3 \text{ dB} . \tag{6.35}$$

Table 6.8 gives TL_{in} values for the duct sizes listed in Table 6.7.

EXAMPLE 6.3

Determine the breakout and breakin sound power of a flat oval duct given the following information: major axis — 24 in.; minor axis — 6 in.; length — 20 ft. The duct is constructed of 24 gauge sheet metal.

SOLUTION

q = mass/unit area of 24 ga. sheet metal = 1.0 lb_m/ft^2 .

Table 6.7 TL_{out} vs. Frequency for Various Flat Oval Ducts

Size — in x in	Gauge	63	125	250	500	1000	2000	4000
12 x 6	24 ga.	31	34	37	40	43	—	—
24 x 6	24 ga.	24	27	30	33	36	—	—
24 x 12	24 ga.	28	31	34	37	—	—	—
48 x 12	22 ga.	23	26	29	32	—	—	—
48 x 24	22 ga.	27	30	33	—	—	—	—
96 x 24	20 ga.	22	25	28	—	—	—	—
96 x 48	18 ga.	28	31	—	—	—	—	—

1/1 Octave Band Center Frequency — Hz

Table 6.8 TL_{in} vs. Frequency for Various Flat–Oval Ducts

Size — in x in	Gauge	63	125	250	500	1000	2000	4000
12 x 6	24 ga.	18	18	22	31	40	—	—
24 x 6	24 ga.	17	17	18	30	33	—	—
24 x 12	24 ga.	15	16	25	34	—	—	—
48 x 12	22 ga.	14	14	26	29	—	—	—
48 x 24	22 ga.	12	21	30	—	—	—	—
96 x 24	20 ga.	11	22	25	—	—	—	—
96 x 48	18 ga.	19	28	—	—	—	—	—

1/1 Octave Band Center Frequency — Hz

Breakout:

$$A_i = 6 \cdot (24 - 6) + \frac{\pi (6)^2}{4} = 136.27 \text{ in.}^2$$

$$12 \cdot 20 \cdot [2 \cdot (24 - 6) + \pi \cdot 6] = 13{,}163.9 \text{ in.}^2 .$$

$$P = 2 \cdot (24 - 6) + \pi \cdot 6 = 54.85 \text{ in.}$$

$$\sigma = \frac{1}{1 + \dfrac{\pi \cdot 6}{2(24 - 6)}} = 0.6563 .$$

$$TL_{out}(\min) = 10 \log_{10} \left[\frac{13{,}163.9}{136.27}\right] = 19.8 \text{ dB} .$$

$$f_L = \frac{8115}{6} = 1352.5 \text{ Hz} .$$

The results are shown below.

	63	125	250	500	1000	2000	4000
TL_{out}, dB	24.2	27.2	30.2	33.3	36.3	—	—
$10 \log_{10} [A_o/A_i]$	19.8	19.8	19.8	19.8	19.8		
$L_{W_r} - L_{W_i}$, dB	−4.4	−7.4	−10.4	−13.4	−16.5	—	—

1/1 Octave Band Center Frequency — Hz

Breakin:

$$f_1 = \frac{6764}{(24 - 6) \cdot \left[1 + \dfrac{\pi \cdot 6}{2(24-6)}\right]^{1/2}} = 304.4 \text{ Hz} .$$

The results are tabulated below.

	1/1 Octave Band Center Frequency — Hz						
	63	125	250	500	1000	2000	4000
TL_{out}, dB	24.2	27.2	30.2	33.3	36.3	—	—
Eq. 6.33a	0.5	9.5	18.5	—	—	—	—
Eq. 6.33b	16.8	16.8	16.8	—	—	—	—
	16.8	16.8	16.8	−3.0	−3.0	—	—
TL_{in}, dB	16.8	16.8	18.5	30.3	33.3	—	—
	3.0	3.0	3.0	3.0	3.0	—	—
$L_{W_t} - L_{W_i}$, dB	−19.8	−19.8	−21.5	−33.3	−36.3	—	—

6.5 INSERTION LOSS OF EXTERNAL LAGGING ON RECTANGULAR DUCTS

External acoustic lagging is often applied to rectangular duct work to reduce the transmission of sound energy from within the duct to surrounding areas. The lagging usually consists of a layer of soft, flexible, porous material, such as fiberglass, covered with an outer impervious layer (Figure 6.5). A relatively rigid material, such as sheet metal or gypsum board, or a limp material, such as sheet lead or loaded vinyl, can be used for the outer covering.

With respect to the insertion loss of externally lagged rectangular ducts, different techniques must be used for rigid and limp outer coverings (11,19). When rigid materials are used for the outer covering, a pronounced resonance effect between the duct walls and the outer covering usually occurs. With limp materials, the variation in the separation between the duct and its outer covering dampens the resonance so that it no longer occurs. For both techniques, it is necessary to determine the low–frequency insertion loss, IL(lf) (dB). It is given by

$$IL(lf) = 20 \log_{10} \left[1 + \frac{M_2}{M_1} \frac{P_1}{P_2} \right] \tag{6.36}$$

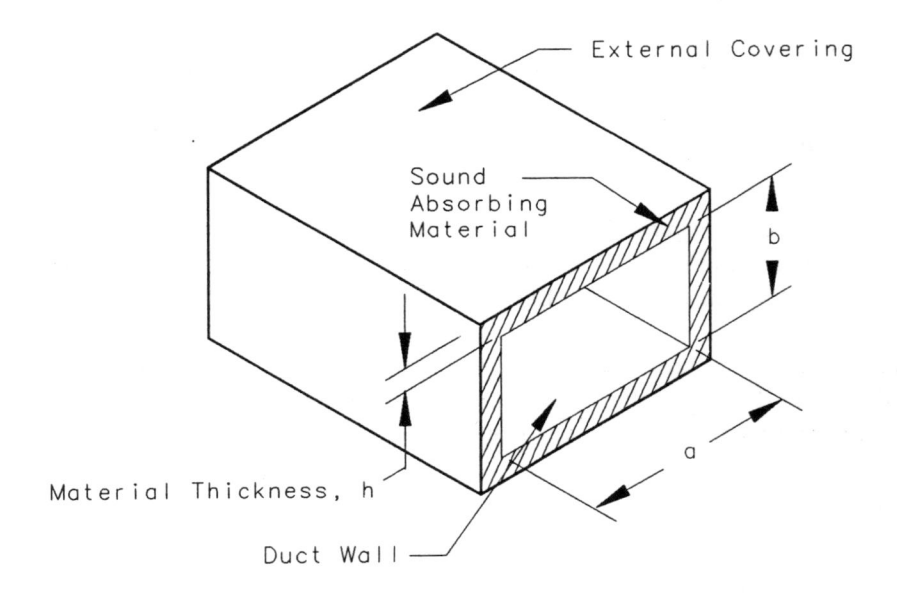

Figure 6.5 External Duct Lagging on Rectangular Ducts

where P_1 is the perimeter of the duct (in.), P_2 is the perimeter of the outer covering (in.), M_1 is the mass per unit area of the duct (lb/ft^2), and M_2 is the mass per unit area of the outer covering (lb/ft^2). P_1 and P_2 are specified by

$$P_1 = 2 \cdot (a + b) \tag{6.37}$$

$$P_2 = 2 \cdot (a + b + 4h) \tag{6.38}$$

where a is the duct width (in.), b is the duct height (in.), and h is the thickness (in.) of the soft, flexible, porous material between the duct wall and the outer covering.

If a rigid outer covering is used, it is necessary to determine the resonance frequency, f_r (Hz), associated with the interaction between the duct wall and outer covering. f_r is given by

$$f_r = 156 \cdot \left[\left[\frac{P_2}{P_1} + \frac{M_2}{M_1} \right] \cdot \frac{P_1}{M_2} \cdot S \right]^{1/2} \tag{6.39}$$

where M_1, M_2, P_1, and P_2 are as previously defined. S is the cross-sectional area (in.^{2}) of the absorbent material and is given by

$$S = 2h \cdot (a + b + 2h) . \tag{6.40}$$

The following procedures for determining the insertion loss for external duct lagging should be used for rigid and limp outer coverings.

Rigid Covering Materials If 1/3 octave band values are desired, draw a line from point B (0.71 f_r) to point A (f_r) on Figure 6.6a. The difference in IL (dB) between points B and A is 10 dB. The equation for this line is

$$IL = IL(lf) - 67.23 \log_{10} \left[\frac{f}{0.71 \cdot f_r} \right] . \tag{6.41}$$

Next, draw a line from point A (f_r) to point C (1.41 f_r) on Figure 6.5a. The equation for this line is

$$IL = IL(lf) - 10 + 67.02 \log_{10} \left[\frac{f}{f_r} \right] . \tag{6.42}$$

From point C (1.41 $\cdot f_r$), draw a line with a slope of 9 dB/octave. The equation for this line is

$$IL = IL(lf) + 29.90 \log_{10} \left[\frac{f}{1.41 \cdot f_r} \right] . \tag{6.43}$$

If 1/1 octave band values are desired, use Equation (6.36) for the 1/1 octave bands below the one that contains f_r. For the 1/1 octave band that contains f_r, subtract 5 dB from IL(lf) obtained from Equation (6.36). For the 1/1 octave bands above the one that contains f_r, use Equation (6.43).

Limp Covering Materials Since there is no pronounced resonance with limp covering materials, the low-frequency insertion loss, IL(lf), is valid up to f_r, after which the insertion loss increases at a rate of 9 dB per octave (Figure 6.6b). For frequencies above f_r, the equation for insertion loss is

$$IL = IL(lf) + 29.90 \log_{10} \left[\frac{f}{f_r} \right] . \tag{6.44}$$

The insertion loss of duct lagging probably does not exceed 25 dB.

(a) Rigid Outer Covering

(b) Limp Outer Covering

Figure 6.6 Insertion Loss Associated with Rectangular External Duct Lagging

According to Cummings, the insertion loss predictions using the procedures described above should be fairly accurate up to about 1,000 Hz for most ducts (11,13). Work reported by Harold indicates that insertion loss values obtained by the above procedures are usually higher than the corresponding values obtained in actual installations (19). Duct lagging may not be a particularly effective method for reducing low-frequency (<100 Hz) duct sound breakout. A more effective method for reducing duct breakout is the use of circular ducts, which have a high transmission loss at low frequencies (1).

EXAMPLE 6.4

Determine the 1/1 octave band insertion loss associated with the external lagging of a rectangular sheet metal duct with the following characteristics: duct dimensions — 8 in. by 8 in.; duct constructed of 18 gauge sheet metal; thickness of absorbent material — 1 in.; outer covering — 1/2 in Gypsum board.

SOLUTION

M_1 = mass/unit area of 18 gauge sheet metal = 2.0 lb/ft^2.

M_2 = mass/using area of one sheet of 1/2 in Gypsum board = 2.1 lb/ft^2.

$P_1 = 2 \cdot (8 + 8) = 32$ in. $P_2 = 2 \cdot (8 + 8 + 4 \cdot 1) = 40$ in.

Thus,

$$IL(lf) = 20 \cdot \log_{10}\left[1 + \frac{2.1}{2.0} \cdot \frac{32}{40}\right] = 5.3 \text{ dB} .$$

$$S = 2 \cdot 1 \cdot (8 + 8 + 2 \cdot 1) = 36 \text{ in.}^2 .$$

The resonance frequency is

$$f_r = 156 \cdot \left[\left[\frac{40}{32} + \frac{2.1}{2.0}\right] \cdot \frac{32}{2.1} \cdot 36\right] = 154 \text{ Hz}$$

where 154 Hz is in the 125 Hz 1/1 octave band. Equation (6.43) can be written as

$$IL = IL(lf) + 29.90 \cdot \log_{10}\left[\frac{f}{1.41 \cdot 154}\right]$$

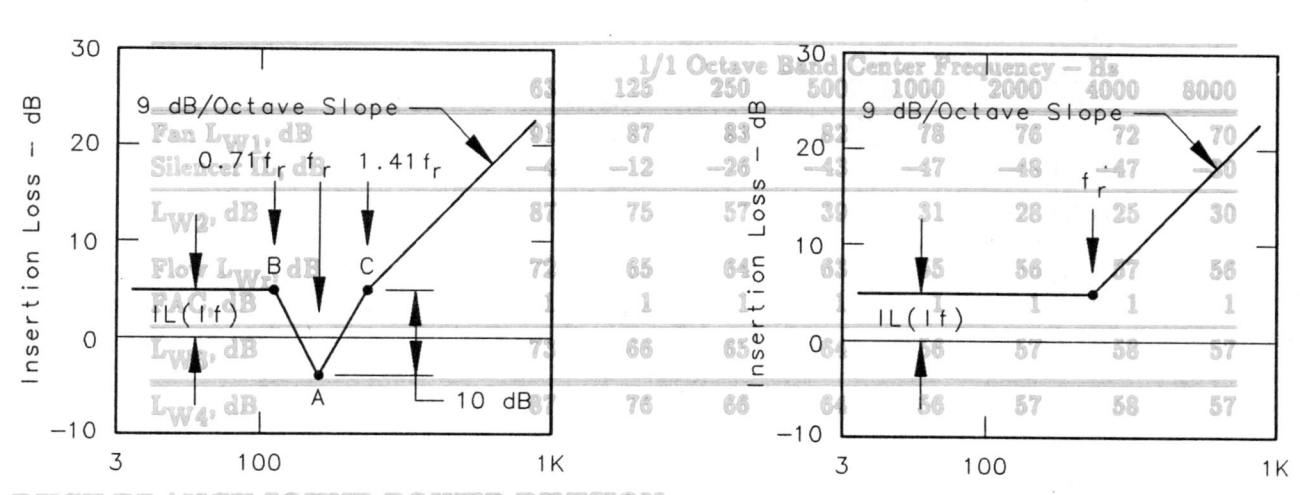

The results are summarized below.

	63	125	250	500	1000	2000	4000	8000
IL(lf), dB	5.3	5.3 −5.0						
IL (9dB/octave), dB			7.1	16.1	25.1			
Max IL value, dB					25.0	25.0	25.0	25.0
IL, dB	5.3	0.3	7.1	16.1	25.0	25.0	25.0	25.0

1/1 Octave Band Center Frequency — Hz

SOUND TRANSMISSION IN INDOOR AND OUTDOOR SPACES — CHAPTER 7

7.1 SOUND TRANSMISSION THROUGH CEILING SYSTEMS

When mechanical equipment is located in the ceiling plenum above an occupied room, noise transmission through the ceiling can be high enough to cause excessive noise levels in that room. Since there are no standard tests for determining the transmission loss through ceiling construction, data are limited. Table 7.1 gives single-pass transmission loss values for various ceiling materials (35,36).

The single-pass transmission loss values in Table 7.1 are for ceilings in which there are no penetrations or acoustical flanking. The acoustical integrity of ceilings can be greatly compromised by these factors. When leaks and/or flanking paths are present, the transmission loss of a ceiling can be significantly reduced. Equation (7.1) gives the corrected transmission loss, TL (dB), taking into account flanking transmission and acoustical leaks,

$$\text{TL} = -10 \cdot \log_{10} \left[(1-\tau) \cdot 10^{(-\text{TL}/10)} + \tau \right] \tag{7.1}$$

where τ is the correction coefficient for type of ceiling. Table 7.2 gives approximate values for τ for various types of ceilings. The values were selected because they yield transmission loss values that agree reasonably well with expected values.

EXAMPLE 7.1

Determine the transmission loss of a ceiling with one layer of 1/2-in. gypsum board that has a surface weight of 2.1 lb/ft^2. The ceiling has few diffusers and the penetrations are well sealed.

SOLUTION

The following transmission loss values are obtained from Table 7.1:

Table 7.1 Transmission Loss Values for Ceiling Materials

Thickness in	Surface Weight lb/ft^2	63	125	250	500	1000	2000	4000
\multicolumn{9}{c}{Gypsum Board}
3/8	1.6	6	11	17	22	28	32	24
1/2	2.1	9	14	20	24	30	31	27
5/8	2.7	10	15	22	26	31	28	30
1	4.6	13	18	26	30	30	29	37

Transmission Loss — dB, 1/1 Octave Band Center Frequency — Hz

Dimensions ftxftxin								
\multicolumn{9}{c}{Acoustical Ceiling Tile — Exposed T-Bar Grid Suspended Lay-in-Ceilings}
2x4x5/8	0.6 – 0.7	4	9	9	14	19	24	26
2x4x1-1/2	0.9	4	9	10	11	15	20	25
2x4x5/8	0.95 – 1.1	5	11	13	15	21	24	28
2x2x5/8	1.2 – 1.3	5	10	11	15	19	22	24
\multicolumn{9}{c}{Acoustical Ceiling Tile — Concealed Spline Suspended Ceiling}
1x1x5/8	1.2	6	14	14	18	22	27	30
\multicolumn{9}{c}{ASHRAE 1987 General Ceiling Tile Insertion Loss Values}
		1	2	4	8	9	9	14

Table 7.2 τ for Different Types of Ceiling Configurations

Ceiling Configuration	τ
Gypboard: No Ceiling Diffusers or Penetrations	0.0001
Gypboard: Few Ceiling Diffusers and Penetrations Well Sealed	0.001
Lay–in–Suspended Tile: No Integrated Lighting of Diffuser System	0.001
Lay–in–Suspended Tile: Integrated Lighting and Diffuser System	0.03

	\multicolumn{7}{c}{1/1 Octave Band Center Frequency – Hz}
	63	125	250	500	1000	2000	4000
TL, dB	9	14	20	24	30	31	27

The correction factor for a suspended ceiling with few ceiling diffusers and well–sealed penetrations is 0.001. Thus,

$$\text{TL} = -10 \log_{10} \left[(1 - 0.001) \cdot 10^{(-\text{TL}/10)} + 0.001 \right].$$

The results are tabulated below.

	\multicolumn{7}{c}{1/1 Octave Band Center Frequency – Hz}
	63	125	250	500	1000	2000	4000
TL, dB	9.0	13.9	19.6	23.0	27.0	27.5	25.2

7.2 RECEIVER ROOM SOUND CORRECTIONS

The sound pressure levels associated with a sound source that occur at a given point in a room depend on the source strength, the acoustical characteristics of the room (surface treatments, furnishings, etc.), the room volume, and the distance of the sound source from the point of observation. There are two types of sound sources associated with HVAC systems that can exist in a room: point source and line source. The point source is usually associated with sound radiated from supply and return air diffusers, equipment items (such as fan–powered terminal units above a lay–in ceiling), and other similar items. The line source is associated with duct breakout noise. Two equations are proposed for use with point sound sources: Thompson's equation and Schultz's equation (33,42,43). Classic diffuse room theory predicts that as a receiver is moved away from a sound source, the sound pressure levels associated with the sound source decrease at the rate of 6 dB for every doubling of distance from the sound source until they reach a constant level, which does not change as the receiver continues to be moved away from the source. Both Thompson and Schultz have reported that this does not happen in most rooms of interest to HVAC engineers. Both indicate that the sound pressure levels, in fact, never reach a constant level. Thompson has reported that the sound pressure levels close to the sound source decrease at a rate of 6 dB per doubling of distance and transition into a rate of decrease of 3 dB per doubling of distance as the distance from the sound source is continually increased. Schultz's work indicates that the sound pressure levels decrease at a rate of 3 dB per doubling of distance from the sound source.

Thompson's equation is based on experimental results, and it is an empirical modification of the classic diffuse room equation (42,43). The equation is

$$L_p = L_W + 10 \log_{10} \left[\frac{Q \, e^{-mr}}{4 \pi r^2} + \frac{MFP}{r} \cdot \frac{4}{R} \right] + 10 \log_{10} [N] + 10.5 \tag{7.2}$$

where Q is the directivity factor associated with the sound source in the direction of the receiver, r is the distance (ft) between the sound source and receiver, MFP is the mean free path (ft), R is the room constant (ft^2), m is the air absorption coefficient (1/ft), and N is the number of point sound sources. Q equals 1 for whole space, 2 for half space, 4 for quarter space, and 8 for eighth space. For most problems of interest to an HVAC designer, sound from a sound source generally radiates into half space; thus, Q will usually equal 2. The mean free path, MFP, is given by

$$MFP = \frac{4 \cdot V}{S} \tag{7.3}$$

where S is the total surface area of the room (ft^2) and V is the total room volume (ft^3). The room constant is specified by

$$R = \frac{S \, \overline{\alpha}_T}{1 - \overline{\alpha}_T} \tag{7.4}$$

where $\overline{\alpha}_T$ is the average room absorption coefficient. $\overline{\alpha}_T$ is given by

$$\overline{\alpha}_T = \overline{\alpha} + \frac{4mV}{S} \tag{7.5}$$

where $\overline{\alpha}$ is the average room absorption coefficient associated with the acoustic characteristics of the room surfaces and 4mV/S is the contribution to room sound absorption associated with air absorption. Table 7.3 gives typical values of $\overline{\alpha}$ for rooms with the specified acoustic characteristics. The values of $\overline{\alpha}$ in Table 7.3 were calculated assuming a 20 ft by 40 ft by 8 ft room. The walls of the room were constructed of a single layer of 5/8-in. gypsum wallboard. The ceiling, floor, furnishings, and occupancy of the room were varied as indicated in Table 7.3, and the corresponding values for average sound absorption coefficients were calculated. The room surface treatments were selected to yield average absorption coefficients in the 500 Hz 1/1 octave frequency band that corresponded to values given in Noise and Vibration Control for live, medium live, average, medium dead, and dead rooms (3). Table 7.4 gives typical values of m. Air absorption associated with values of m need only be considered in large rooms, such as auditoriums and large, open office areas.

Schultz's equation for converting from L_W to L_p in a receiver room is an empirical equation based solely on experimental results. Schultz found that the sound pressure levels in a receiver room are functions of the sound power level of the sound source, the room volume, and 1/1 octave band center frequency (33). For individual sound sources, the sound pressure level, L_p (dB), in a receiver room is given by

$$L_p = L_W - 10 \log_{10} [r] - 5 \log_{10} [V] - 3 \log_{10} [f] + 10 \log_{10} [N] + 25 \tag{7.6}$$

where L_W is the sound power level (dB) of the sound source, r is the distance (ft) between the sound source and receiver, V is the room volume (ft^3), f is the 1/1 octave band center frequency (Hz), and N is the number of sound sources. For an array of distributed ceiling diffusers, Equation (7.6) can be expressed as

$$L_p(5 \text{ ft}) = L_{Ws} - 27.6 \log_{10} [h] - 5 \log_{10} [X] - 3 \log_{10} [f] + 1.3 \log_{10} [N] + 30 \tag{7.7}$$

where L_p(5 ft) is the sound pressure level (dB) at a distance 5 feet above the floor, L_{Ws} is the sound power level (dB) associated with a single diffuser in the array, h is the ceiling height, N is the number of ceiling diffusers, and X is the ratio of the floor area served by each diffuser divided by the square of the ceiling height (X = 1 if the area served equals h^2).

Caution should be exercised in using both Thompson's and Schultz's equations. With regard to Thompson's equation, it is necessary to make an assumption as to the acoustic characteristics of the room being analyzed. Most rooms of interest to the HVAC designer will be acoustically medium dead. Thus, the average absorption coefficient

Table 7.3 Average Sound Absorption Coefficients, $\bar{\alpha}$, for Typical Receiving Rooms

Type of Room	\multicolumn{7}{c}{1/1 Octave Band Center Frequency — Hz}
	63	125	250	500	1000	2000	4000
Dead Acoustical ceiling, plush carpet, soft furnishings, people	0.26	0.30	0.35	0.4	0.43	0.46	0.52
Medium Dead Acoustical ceiling, commercial carpet, people	0.24	0.22	0.18	0.25	0.30	0.36	0.42
Average Acoustical ceiling or commercial carpet, people	0.25	0.23	0.17	0.20	0.24	0.29	0.34
Medium Live Some acoustical material, people	0.25	0.23	0.15	0.15	0.17	0.20	0.23
Live People only	0.26	0.24	0.12	0.10	0.09	0.11	0.13

Table 7.4 Air Absorption Coefficients

	63	125	250	500	1000	2000	4000
m (1/ft)	0	0	0	0	0	0.0009	0.0029

in Table 7.3 for a medium dead room should be selected for most calculations. It is generally understood that the conversion from L_W to L_p in a receiving room is a function of room sound absorption, typically described by the average absorption coefficient of the receiving room. In Schultz's equations, the effects of room absorption are contained in the room volume term. Equations (7.6) and (7.7) apply primarily to rooms that are acoustically medium dead. These equations will give reasonable results for rooms that have acoustical characteristics that range from average to dead. The lowest frequency band for which data were presented by Thompson was the 250 Hz 1/1 octave band (42). For Schultz, the lowest frequency band was the 125 Hz 1/1 octave band (33). Also, with regard to Schultz's data, there are some questions relative to the value of the calibrated sound power level in the 125 Hz 1/1 octave band of the reference sound source used to generate the data that were utilized in developing his equation. Equation (7.6) is a completely empirical equation. As such, there are currently no data available that can be used to justify extrapolating this equation or Equation (7.7) to the 63 Hz 1/1 octave band. Since Thompson's equation is a modification of the classic diffuse room equation, there is some justification for extrapolating Equation (7.2) to the 63 Hz and 125 Hz 1/1 octave bands. However, when this is done, a great deal of caution should be exercised when interpreting the results in these bands. In the 1/1 octave frequency bands from 125 Hz to 8,000 Hz and for medium dead rooms, Schultz's equation currently gives the most accurate and dependable results.

When considering the sound pressure levels in a room that are associated with duct breakout, the sound source is the duct, which must be considered as a line source. For converting from duct breakout sound power levels to the corresponding sound pressure levels in the room, the equation for a line sound source should be used. It is:

$$L_p = L_W + 10 \log_{10} \left[\frac{Q}{4\pi\, r\, L} + \frac{4}{R} \right] + 10.5 \tag{7.8}$$

where L_W and L_p are defined as before, Q is as specified before, r is the distance (ft) from the line source to the receiver, L is the length of the line source, and R is defined by Equation (7.4). This is the classic diffuse room equation for a line sound source. Currently there is no other information for converting from L_W to L_p for line sound sources.

EXAMPLE 7.2

Determine the values for converting from L_W to L_p for a room that is 15 ft long, 10 ft wide, and 8 ft high. The sound source is a single diffuser located in the ceiling. The distance between the sound source and receiver is 8 feet. Use Equation (7.6)

SOLUTION

$V = 15 \cdot 10 \cdot 8 = 1200 \text{ ft}^3$.

The results are tabulated below.

	\multicolumn{8}{c}{1/1 Octave Band Center Frequency – Hz}
	63	125	250	500	1000	2000	4000	8000
$-10 \log_{10} [8]$	−9.0	−9.0	−9.0	−9.0	−9.0	−9.0	−9.0	−9.0
$-5 \log_{10} [1200]$	−15.4	−15.4	−15.4	−15.4	−15.4	−15.4	−15.4	−15.4
$-3 \log_{10} [f]$	−5.4	−6.3	−7.2	−8.1	−9.0	−9.9	−10.8	−11.7
	25.0	25.0	25.0	25.0	25.0	25.0	25.0	25.0
$L_p - L_W$, dB	−4.8	−5.7	−6.6	−7.5	−8.4	−9.3	−10.2	−11.1

7.3 SOUND TRANSMISSION THROUGH MECHANICAL EQUIPMENT ROOM WALLS, FLOOR, OR CEILING

There are many situations in which the mechanical equipment room is adjacent to an occupied space and it is necessary to determine the sound pressure levels in the occupied space that are associated with sound radiated from equipment in the mechanical equipment room. Often the equation used to make this calculation is

$$NR = TL - 10 \log_{10} \left[\frac{S_W}{a} \right] \tag{7.9}$$

where TL is the transmission loss (dB) of the wall between the mechanical equipment room and the adjacent room, S_W is the surface area (ft^2) of the common wall between the two rooms, a is the total acoustic absorption (ft^2) of the adjacent room, and NR is noise reduction (dB) (27). NR is

$$NR = L_{p(m)} - L_{p(a)} \tag{7.10}$$

where $L_{p(m)}$ is the sound pressure level (dB) in the mechanical equipment room and $L_{p(a)}$ is the sound pressure level (dB) in the adjacent room. Equation (7.9) has some very severe limitations associated with it. With the exception of the vicinity of the sound source, it assumes that the sound fields in the mechanical equipment and adjacent rooms are totally diffuse. That is, the sound pressure levels in both rooms are the same throughout. This may be the case in the mechanical equipment room if it has all hard reflecting surfaces. If the mechanical equipment room has sound–absorbing materials on the walls or other mechanisms that take sound energy out of the room, $L_{p(m)}$ will not be the same throughout the room. Usually, $L_{p(a)}$ in the adjacent room is not constant throughout. When $L_{p(m)}$ and $L_{p(a)}$ are not constant throughout, they are functions of the location in the room where the respective sound pressure levels are to be determined and of the acoustic and geometric characteristics of the respective rooms. Equation (7.9) does not take any of these factors into account. Finally, in many situations relative to the sound levels that exist in a mechanical equipment room, the sound power levels, not the sound pressure levels, of a sound source are given. Equation (7.9) does not give a means of easily converting from sound power levels to typical sound pressure levels within the room.

To determine the sound pressure levels in an adjacent room associated with noise that exits in a mechanical equipment room, it is necessary to develop a series of equations that take into account the sound power levels of the sound source and the acoustic and geometric characteristics of the mechanical equipment room, the transmission loss characteristics of the wall separating the two rooms, and the acoustic and geometric characteristics of the adjacent room. First, consider the mechanical equipment room. To proceed, it is necessary to determine the acoustical intensity, and consequently sound power, that is incident on the common wall between the mechanical equipment room and the adjacent room. There are two types of acoustical power that are incident on the wall: reverberent and direct. The direct field sound power is that sound power that is directly radiated from a sound source before it undergoes its first reflection. The reverberent field sound power is the sound power that is present after the sound waves have undergone their first reflection. Work completed by Reynolds and Bledsoe indicates that the sound power, $L_{W(wall)}$ (dB), incident on a wall can be given by

$$L_{W(wall)} = L_W + 10 \log_{10} \left[S_W \left[\frac{(1 - \overline{\alpha})}{S_M \overline{\alpha}} + \frac{1}{4 S_W + 2\pi l^2} \right] \right] \qquad (7.11)$$

where L_W is the sound power level (dB) of the sound source (usually a fan), S_W is the area of the wall (ft^2), $\overline{\alpha}$ is the average absorption coefficient of the surfaces of the mechanical equipment room, S_M is the total surface area (ft^2) of the mechanical equipment room walls, ceiling and floor, and l is the distance between the sound source and the wall (30). Equation (7.11) assumes the sound source is radiating into half space and the sound source is radiating sound equally in all directions. Many sound sources do not radiate sound equally in all direction. If there is a directional characteristic associated with a particular sound source, Equation (7.11) can be written as

$$L_{W(wall)} = L_W + 10 \log_{10} \left[S_W \left[\frac{(1 - \overline{\alpha})}{S_M \overline{\alpha}} + \frac{DF}{4 S_W + 2\pi l^2} \right] \right] \qquad (7.12)$$

where DF is the directivity factor associated with the sound source in the direction of the wall of interest.

Equation (7.12) can be used to determine the sound power incident on a hole in the mechanical equipment room wall that has an area of S_H (ft^2). For this situation, $S_W = S_H$ and $S_H \ll 2\pi l^2$. Thus, Equation (7.12) becomes

$$L_{W(hole)} = L_W + 10 \log_{10} \left[S_H \left[\frac{(1 - \overline{\alpha})}{S_M \overline{\alpha}} + \frac{DF}{2\pi l^2} \right] \right]. \qquad (7.13)$$

Generally the surfaces of a mechanical equipment room are hard, reflecting surfaces. Table 7.5 gives the sound absorption coefficients, α, of common materials used in mechanical equipment rooms as a function of 1/1 octave band center frequencies (25,27). The ceiling and floor are generally poured concrete. The walls are usually concrete blocks, poured concrete, or gypsum board. Typically the walls are constructed of one type of material. When this is the case, the average sound absorption coefficient for a mechanical equipment room can be calculated from

$$\overline{\alpha} = \frac{S_1 \alpha_1 + S_2 \alpha_2}{S_M} \qquad (7.14)$$

where α_1 is the absorption coefficient of the ceiling and floor, α_2 is the absorption coefficient of the walls, S_1 is the surface area (ft^2) of the ceiling and floor, S_2 is the surface area (ft^2) of the walls, and S_M is the total surface area (ft^2) of the mechanical equipment room. If a fiberglass blanket is added to the surfaces of a mechanical equipment room, the new average absorption coefficient, α_a, can be estimated from

$$\alpha_a = \frac{PC \cdot \alpha_3 + (100 - PC) \cdot \overline{\alpha}}{100} \tag{7.15}$$

where $\overline{\alpha}$ is the absorption coefficient from Equation (7.14), α_3 is the absorption coefficient of the fiberglass blanket, and PC is the percent of the total room surface that is covered by the fiberglass blanket.

After the sound power levels incident on the common wall between a mechanical equipment room and an adjacent room have been calculated, it is necessary to determine the transmission loss through the wall. Tables 7.6 through 7.8 give transmission loss, TL (dB), values for wall configurations that are commonly used for mechanical equipment rooms (9). The transmission loss values given in Tables 7.6 through 7.8 were obtained under ideal

Table 7.5 Sound Absorption Coefficients for Various Construction Materials

Material	\multicolumn{7}{c}{1/1 Octave Band Center Frequency — Hz}
	63	125	250	500	1000	2000	4000
1. Poured Concrete	0.01	0.01	0.01	0.02	0.02	0.02	0.03
2. Concrete Block: Painted	0.11	0.10	0.05	0.06	0.07	0.09	0.08
3. Concrete Block: Unpainted	0.17	0.18	0.22	0.15	0.15	0.19	0.12
4. Single Layer 5/8" Gypsum Board	0.31	0.29	0.10	0.06	0.05	0.05	0.05
5. Double Layer 5/8" Gypsum Board	0.13	0.10	0.05	0.05	0.05	0.05	0.05
6. 2" 3 pcf Fiberglass Insulation	0.15	0.22	0.82	1.00	1.00	1.00	1.00
7. 3" 3 pcf Fiberglass Insulation	0.48	0.53	1.00	1.00	1.00	1.00	1.00
8. 4" 3 pcf Fiberglass Insulation	0.76	0.84	1.00	1.00	1.00	1.00	0.97

Table 7.6 Transmission Loss Values of Drywall Configurations

Wall Configuration	\multicolumn{7}{c}{1/1 Octave Band Center Frequency, Hz}
	63	125	250	500	1000	2000	4000
1. 3 5/8" metal studs; 5/8" gypsum wallboard screwed to both sides of studs; all joints taped and finished	11	20	30	37	47	40	44
2. Same as (1) except 2 in 3 lb/cu. ft. fiberglass blanket place between studs	14	23	40	45	53	47	48
3. 3 5/8" metal studs; 2 layers 5/8" gypsum wallboard screwed to both sided of studs; all joints taped and finished	19	27	40	46	52	48	48
4. Same as (3) except 3 in 3lb/cu ft fiberglass blanket placed between studs	24	32	43	50	52	49	50
5. 2 1/2" metal studs; 2 layers 5/8" gypsum wallboard screwed to both sides of studs; all exposed joints taped and finished	18	26	37	45	51	47	49
6. Same as (5) except 2" 3 pcf fiberglass blanket placed between studs	22	31	44	50	53	49	51
7. 3 5/8" metal studs; 2 Layers 5/8" gypsum wallboard screwed to horizontal resilient channels on one side; 3 layers 5/8" gypsum wallboard screwed to studs on other side; 3 in 3 pcf fiberglass between studs	27	40	54	61	65	63	67

Table 7.7 Transmission Loss Values of Masonry/Floor/Ceiling Configurations

Wall Configuration	63	125	250	500	1000	2000	4000
1. 6x8x18 in 3–Cell Light Concrete Block; 21 lb/Block	24	28	34	39	44	50	53
2. 6x8x18 in 3–Cell Dense Concrete Block; 36 lb/Block	27	31	36	42	49	50	56
3. 8x8x18 in 3–Cell Light Concrete Block; 28 lb/Block	31	33	37	40	46	51	55
4. 8x8x18 in 3–Cell Light Concrete Block; 34 lb/Block	29	33	39	45	49	56	60
5. 8x8x18 in 3–Cell Dense Concrete Block; 48 lb/Block	30	34	40	49	52	59	57
6. 3 in thick 160 pcf Concrete; 40 lb/sq ft	30	35	39	43	52	58	64
7. 4 in thick 144 pcf Concrete; 48 lb/sq ft	32	34	35	37	42	49	55
8. 5 in thick 155 pcf Concrete 65 lb/sq ft	37	41	43	47	54	59	63
9. 6 in thick 160 pcf Concrete; 80 lb/sq ft	33	38	43	50	58	64	68
10. 8 in thick 144 pcf Concrete; 96 lb/sq ft	40	44	47	54	58	63	67
11. Prefabricated 3 in Deep trapezoidal Concrete Channel Slabs Mortared together on 20 in centers; 28 lb/sq ft	33	34	34	38	45	55	61

Table 7.8 Transmission Loss Values of Painted Masonry Block Walls and Painted Block Walls with Resiliently Mounted Gypsum Wallboard

Wall Configuration	63	125	250	500	1000	2000	4000
1. 6x8x18 in 3–Cell Light Concrete Block; 21 lb/Block Both Sides Painted	32	36	36	41	45	54	58
2. 6x8x18 in 3–Cell Dense Concrete Block; 36 lb/Block; Both Sides Painted	32	36	38	42	49	53	59
3. 8x8x18 in 3–Cell Light Concrete Block; 28 lb/Block; Both Sides Painted	34	38	38	40	46	54	58
4. Same as (1) Except 1/2" Gypsum Wallboard screwed to resilient channels on one side	32	36	43	48	61	66	66
5. Same as (3) Except 5/8" Gypsum Wallboard screwed to resilient channels on one side	35	39	40	48	59	61	66

Table 7.9 Correction Factors for the Quality of Construction

Quality of Construction	τ
Excellent — No acoustic leaks or penetrations	0.00001
Good — Very few acoustic leaks and penetrations	0.0001
Average — Many acoustic leaks and penetrations	0.001
Poor — Many acoustic leaks and visible holes	0.01

laboratory conditions. These conditions rarely exist in the field. Acoustic leaks and flanking can substantially reduce these transmission loss values. Equation (7.16) gives corrected transmission loss values taking into account flanking transmission and acoustic leaks,

$$\text{TL} = -10 \log_{10}\left[(1-\tau)\cdot 10^{-\text{TL}/10} + \tau\right] \tag{7.16}$$

where τ is the correction coefficient associated with the quality of construction. Table 7.9 gives specified values of τ for different qualities of construction. The values were selected because they yield transmission loss values that agree reasonably well with expected values.

When both the sound power levels, $L_{W(\text{wall})}$, associated with the acoustic intensity incident on the common wall between a mechanical equipment room and an adjacent room and the transmission loss, TL, of the wall have been determined, the sound power levels, $L_{W(\text{room})}$, associated with the acoustic energy radiated into the adjacent room are obtained by subtracting the transmission loss of the wall from the sound power levels incident on the wall, or

$$L_{W(\text{room})} = L_{W(\text{wall})} - \text{TL}. \tag{7.17}$$

The final step is to convert from the sound power levels, $L_{W(\text{room})}$, of the sound energy radiated into the receiving room to the corresponding sound pressure levels, $L_{p(\text{room})}$. In the vicinity of the wall, the sound pressure levels will remain fairly constant as distance from the wall is increased. For this case, the sound pressure levels are given by

$$L_{p(\text{room})} = L_{W(\text{room})} + 10 \log_{10}\left[\frac{1}{S_W} + \frac{4(1-\alpha)}{S_R \overline{\alpha}}\right] + 10.5 \tag{7.18}$$

where $\overline{\alpha}$ is the average absorption coefficient for the receiving room, S_R is the total surface area (ft^2) of the receiving room, and S_W is the area (ft^2) of the common wall between the mechanical equipment room and the adjacent (receiving) room (2). Typical values for the average absorption coefficient for the receiving room are given in Table 7.3. As distance from the wall increases, the sound pressure levels begin to decrease (33,42). This usually occurs when the surface area of a hemisphere with a radius equal to the distance from the wall becomes greater than S_W. It is assumed that the sound is radiating into half space from the wall. For this case, the sound pressure levels are given by

$$L_{p(\text{room})} = L_{W(\text{room})} + 10 \log_{10}\left[\frac{1}{2\pi r^2} + \frac{\text{MFP}}{r}\cdot\frac{4}{R}\right] + 10.5 \tag{7.19}$$

where r is the distance (ft) from the wall, MFP is given by Equation (7.3), and R is given by equation (7.4). If the distance, r, from the wall at which $L_{p(\text{room})}$ is to be calculated is such that $2\pi r^2 \leq S_W$, use Equation (7.18). If $2\pi r^2 > S_W$, use Equation (7.19).

EXAMPLE 7.3

A mechanical equipment room has the following dimensions: height = 12.75 ft, length = 12 ft, and width = 14.5 ft. An adjacent room has the following dimensions: height = 9.0 ft, length = 12 ft, and width = 14.5 ft. The longer wall is the common wall between the two rooms. It has the following dimensions: height = 9.0 ft and width = 14.5 ft. A sound source is placed on the floor of the mechanical equipment room 3 ft from the common wall. The receiving point in the adjacent room is 6 feet from the common wall. The mechanical equipment room is constructed of painted concrete blocks. There is no fiberglass sound absorbing material in the mechanical equipment room. The common wall is: 8 in. by 8 in. by 18 in. three-cell light concrete block (28 lb/block) painted on both sides. Assume the adjacent room is acoustically medium dead and the construction of the wall is excellent. Determine the values of the sound pressure levels in the receiving room, $L_{p(\text{room})}$, minus the value of the sound power levels of the sound source in the mechanical equipment room, L_W. Assume DF = 1.

7–10 SOUND TRANSMISSION IN INDOOR AND OUTDOOR SPACES

SOLUTION

$S_1 = 2 \cdot 12 \cdot 14.5 = 348 \text{ ft}^2$.

$S_2 = 2 \cdot (12.75 \cdot 12 + 12.75 \cdot 14.5) = 675.75 \text{ ft}^2$.

$S_M = 675.75 + 348 = 1{,}023.75 \text{ ft}^2$.

$S_W = 9 \cdot 14.5 = 130.5 \text{ ft}^2$.

$\tau = .00001$ (Table 7.9).

$S_R = 2 \cdot (9 \cdot 12 + 9 \cdot 14.5 + 12 \cdot 14.5) = 825 \text{ ft}^2$.

V = Volume of adjacent room = $9 \cdot 12 \cdot 14.5 = 1566 \text{ ft}^3$.

$2 \cdot \pi \cdot r^2 = 2 \cdot \pi \cdot 6^2 = 226.2 \text{ ft}^2$.

$2\pi r^2 > S_W$; therefore, use Equation (7.19) for calculating $L_{p(\text{room})}$.

MFP = $4 \cdot 1566 / 825 = 7.6 \text{ ft}$.

The results are tabulated below.

	\multicolumn{7}{c}{1/1 Octave Band Center Frequency — Hz}						
	63	125	250	500	1000	2000	4000
Mechanical equipment room:							
$\overline{\alpha}$, eq. (7.14)	.076	.069	.036	.046	.053	.066	.063
$(1-\overline{\alpha})/S_M \overline{\alpha}$	.0119	.0131	.0262	.0202	.0175	.0138	.0145
$1/(4S_w + 2\pi l^2)$	.0017	.0017	.0017	.0017	.0017	.0017	.0017
$L_{W(\text{wall})} - L_W$, dB	2.4	2.9	5.6	4.6	4.0	3.1	3.3
TL through common wall:							
TL of wall (Table 7.8)	34.0	38.0	38.0	40.0	46.0	54.0	58.0
TL, eq. (7.16)	−33.9	−37.7	−37.7	−39.6	−44.5	−48.5	−49.4
Adjacent room:							
$\overline{\alpha}$ (Table 7.3)	0.24	0.22	0.18	0.25	0.30	0.36	0.42
$(\text{MFP}/r) \cdot [4(1-\overline{\alpha})/S_R \overline{\alpha}]$	.0194	.0218	.0280	.0184	.0143	.0106	.0077
$1/2\pi r^2$	.0044	.0044	.0044	.0044	.0044	.0044	.0044
$L_{p(\text{rm})} - L_{W(\text{rm})}$, dB	−5.7	−5.3	−4.4	−5.9	−6.8	−7.7	−8.6
$L_{p(\text{rm})} - L_W$, dB	−37.2	−40.1	−36.5	−40.9	−47.3	−53.1	−54.7

EXAMPLE 7.4

The long wall in the mechanical equipment room in Example 1 has a 48 in. by 36 in. hole in it. The hole is 6 ft from the sound source. Determine the change in sound power levels between the sound source and the hole. Assume DF = 1.

SOLUTION

From Example 1:

$S_1 = 348 \text{ ft}^2 \quad S_2 = 675.75 \text{ ft}^2 \quad S_M = 1023.75 \text{ ft}^2$.

S_H = Area of hole in wall = $3 \cdot 4 = 12 \text{ ft}^2$.

The results are tabulated below.

	1/1 Octave Band Center Frequency — Hz						
	63	125	250	500	1000	2000	4000
$\bar{\alpha}$, eq. (7.14)	.076	.069	.036	.046	.053	.066	.063
$(1-\bar{\alpha})/S_M\bar{\alpha}$	.0119	.0131	.0262	.0202	.0175	.0138	.0145
$1/2\pi l^2$	.0044	.0044	.0044	.0044	.0044	.0044	.0044
$L_{W(hole)} - L_W$, dB	−7.1	−6.8	−4.4	−5.3	−5.8	−6.6	−6.4

7.4 SOUND TRANSMISSION IN OUTDOOR ENVIRONMENTS

There are times when it is necessary to determine the sound pressure levels associated with HVAC equipment outdoors. When investigating the propagation of sound outdoors, it is necessary to take into account meteorological effects on sound propagation, the attenuation effects of ground coverings, atmospheric sound absorption, the sound attenuation associated with barriers, and the effects of reflecting surfaces. Accurate acoustical analyses associated with all of these can be rather complicated. However, this accuracy often is not justified because the acoustical effects of the above factors are extremely variable. The acoustical analysis presented in this section will be simplified to give an estimate of outdoor sound pressure levels associated with HVAC equipment that are affected by spherical spreading, reflecting surfaces and barriers.

The sound pressure level, L_p (dB), of a simple sound source associated with spherical spreading into whole space (no reflecting surfaces) is

$$L_p = L_W - 20 \log_{10}\left[\frac{r}{r_{ref}}\right] + DI \tag{7.20}$$

where L_W is the sound power level (dB) of the sound source, DI is the directivity index (dB), r is the distance from the source (ft) at which L_p is determined, and r_{ref} is the reference distance equal to .9252 ft. When r_{ref} is in feet, Equation (7.20) becomes

$$L_p = L_W - 20 \log_{10}[r] + DI - 0.7 . \tag{7.21}$$

When considering the propagation of sound over the ground, the ground must be considered as a reflecting plane. If the sound source is located close to the ground, it can be considered to radiate sound into half space. When this is the case, DI equals 3 dB. Thus, Equation (7.21) becomes

$$L_p = L_W - 20 \log[r] + 2.3 . \tag{7.22}$$

If the sound source is near a vertical reflecting surface, the sound pressure level at a distance r from the sound source is given by

$$L_p = L_W - 20 \log_{10}[r] + 2.3 + \Delta L \tag{7.23}$$

where ΔL is the correction term associated with the vertical reflecting surface (20). Figure 7.1 shows a point sound source near a hard reflecting surface. The effect of the reflecting surface can be modeled by placing an image sound source on the side of the reflecting surface opposite the sound source. As the figure indicates, there are two paths between the sound source and the receiver. One is the path of the directly radiated sound wave between the source and receiver. The distance of this path is r_s. The other is the path of the reflected sound wave between the sound source and the receiver. The distance of this path is r_i; r_i also is the distance between the image sound source and the receiver. If the reflecting surface is a vertical surface, the effects of this surface and the corresponding values of ΔL are a function of the distance between the sound source and the reflecting surface and the distance between the sound source and the receiver (20). This relation is shown in Figure 7.2 and is expressed by

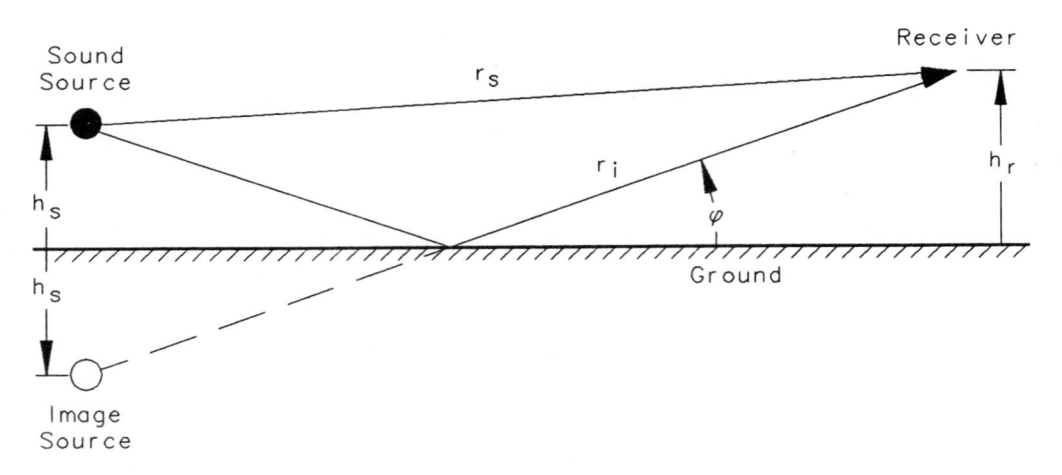

Figure 7.1 Sound Source near a Reflecting Surface

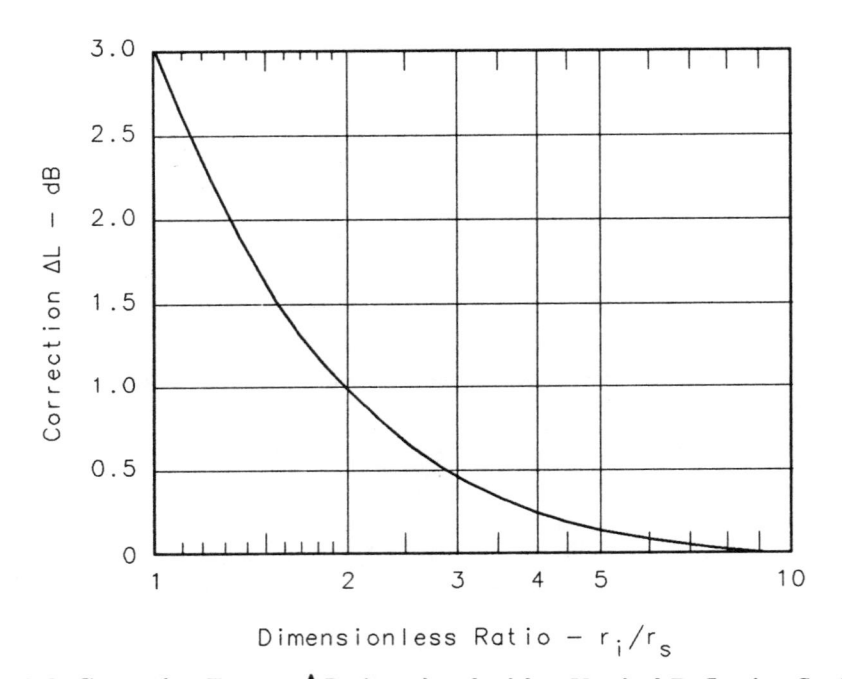

Figure 7.2 Correction Factor, ΔL, Associated with a Vertical Reflecting Surface

$$\Delta L = 3.00 - 9.29 \log_{10}\left[\frac{r_i}{r_s}\right] + 10.13 \left[\log_{10}\left[\frac{r_i}{r_s}\right]\right]^2 - 3.84 \left[\log_{10}\left[\frac{r_i}{r_s}\right]\right]^3 \qquad (7.24)$$

$$\Delta L = 0 \quad \text{for} \quad \frac{r_i}{r_s} > 10. \qquad (7.25)$$

Nonporous barriers, if placed between the sound source and the receiver, can result in significant excess attenuation of sound (Figure 7.3). This is because sound can reach the receiver only by diffraction around or over the boundaries of the barrier. The results presented here are based on experimental data that are consistent with the results of optical–diffraction theory. According to Fresnel diffraction theory, only that portion of a wave field due to a sound source that is incident along the top edge of a barrier contributes appreciably to the wave field that is diffracted over the wall. Figure 7.4 shows the excess attenuation associated with an infinite barrier for the case

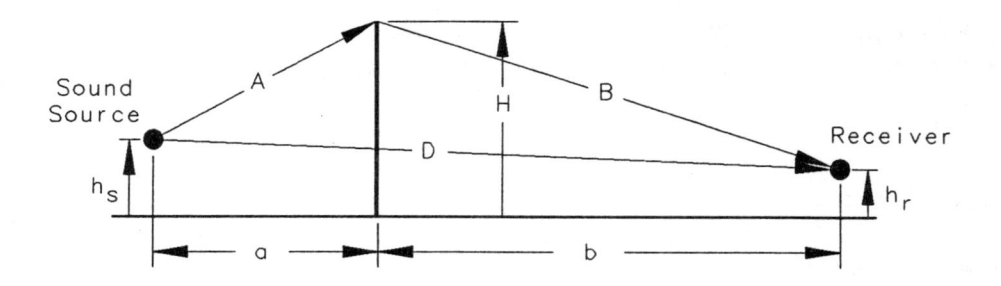

Figure 7.3 Geometry of Sound Propagation Path over or around a Barrier

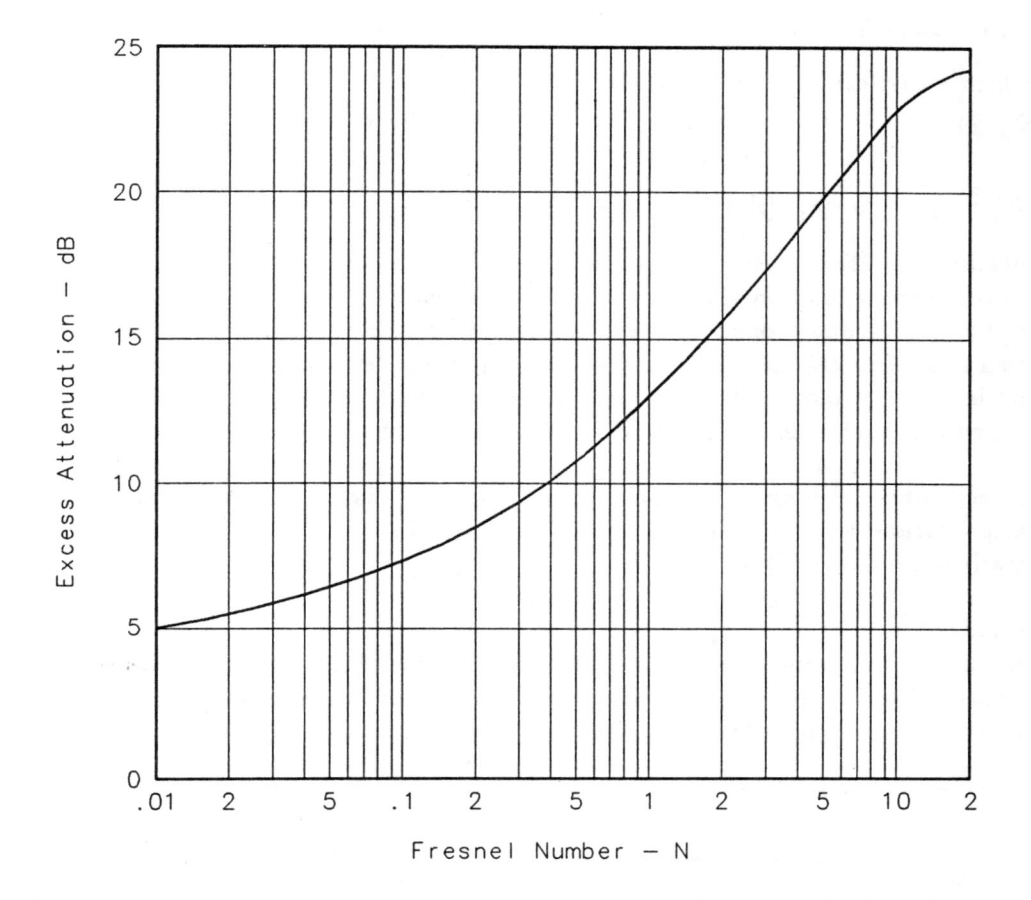

Figure 7.4 Excess Attenuation of Sound Associated with a Rigid Infinite Barrier

where the receiver and/or a point sound source are at some point above the ground (3,17). For a finite sound source, assume the position of the sound source is the geometric center of the source.

The excess attenuation in Figure 7.4 is plotted as a function of the Fresnel number, N, where

$$N = \frac{2}{\lambda} \delta \qquad (7.26)$$

where λ is the wave length (ft) that corresponds to the center of the frequency band being analyzed. δ is called the path length difference (ft) and is given by

$$\delta = A + B - D \qquad (7.27)$$

It is desirable to write Equation (7.27) in terms of the sound source and receiver distances a (ft) and b (ft) from the barrier, sound source and receiver heights h_s (ft) and h_r (ft), and the barrier height H (ft). Thus,

$$A = \left[a^2 + (H - h_s)^2\right]^{1/2} \tag{7.28}$$

$$B = \left[b^2 + (H - h_r)^2\right]^{1/2} \tag{7.29}$$

$$D = \left[(a + b)^2 + (h_s - h_r)^2\right]^{1/2} . \tag{7.30}$$

The excess attenuation, ΔA, can be determined using Figure 7.4 or the following equation

$$\Delta A = 13.24 + 7.80 \cdot \log_{10}[N] + 1.17\left[\log_{10}[N]\right]^2 - 0.37\left[\log_{10}[N]\right]^3 . \tag{7.31}$$

Figure 7.4 indicates that the value of ΔA does not exceed 24 dB. Thus, the value of ΔA in equation (7.31) must be less than or equal to 24 dB. When the excess attenuation for a barrier is taken into account equation (7.23) becomes

$$L_p = L_W - 20\log_{10}[r] + 2.3 + \Delta L - \Delta A . \tag{7.32}$$

Many practical situations exist where barriers are of finite length. If the sound source is a point source, the sound level at the receiver location is obtained by first calculating the individual sound pressure levels for sound traveling over the barrier and around each end of the barrier. After the individual sound pressure levels at the receiver location have been calculated, the overall sound pressure level is obtained by adding the three individual sound pressure levels. For most situations where a barrier is used to attenuate noise from HVAC equipment, the barrier will enclose the equipment on three sides. When this is the case the barrier can be considered an infinite barrier.

With respect to barrier design, the following usually applies. Excess barrier attenuations of 10 dBA or less are easily attainable. Excess barrier attenuations up to 15 dBA are difficult to obtain. Excess barrier attenuations over 20 dBA are nearly impossible to obtain (26).

EXAMPLE 7.5

A fan is located 3 ft from a vertical reflecting surface and 5 ft from a 6-ft barrier. The barrier can be considered infinite. The fan is located 2 ft above the ground and the receiver is located 5 ft above the ground. Determine the sound pressure level 15 ft from the fan if the sound power levels of the fan are as follows.

	63	125	250	500	1000	2000	4000	8000
Fan L_W, dB	91	87	83	82	78	76	72	69

1/1 Octave Band Center Frequency — Hz

SOLUTION

The distance correction for this example is:

$$20\log_{10}[15/0.9252] = 24.2 \text{ dB} .$$

The excess attenuation associated with a barrier is determined using Equations (7.9) through (7.14).

$$A = \left[5^2 + (6-2)^2\right]^{1/2} = 6.4 \text{ ft} .$$

$$B = \left[10^2 + (6-2)^2\right]^{1/2} = 10.8 \text{ ft} .$$

$$D = \left[(5+10)^2 + (2-2)^2\right]^{1/2} = 15 \text{ ft} .$$

The path length difference, δ, is given by

$\delta = A + B - D = 6.4 + 10.8 - 15 = 2.2$ ft.

The Fresnel number, N, is then given by

$$N = \frac{2}{\lambda} \cdot \delta = \frac{2 \cdot f}{c_o} \cdot \delta = \frac{2 \cdot f}{1125} \cdot \delta \qquad N = \frac{f}{562.5} \cdot 2.2 = \frac{f}{255.7}.$$

The results for calculating N are summarized below.

	1/1 Octave Band Center Frequency — Hz								
	63	125	250	500	1000	2000	4000	8000
N	0.13	0.26	0.51	1.03	2.05	4.11	8.22	16.44

The excess barrier attenuation can now be determined using Equation 7.31.

The correction for the vertical reflecting surface can be determined using equation 7.24. $r_s = 15$ and $r_i = 15 + 2 \cdot 3$.

$$\log_{10}\left[\frac{21}{15}\right] = 0.1461.$$

Thus,

$$\Delta L = 3.00 - 9.29 \cdot (0.1461) + 10.13 \cdot (0.1461)^2 - 3.84 \cdot (0.1461)^3 = 1.9 \text{ dB}.$$

Since the fan is reflecting into half space, DI = 3 dB. The results are summarized below.

	1/1 Octave Band Center Frequency — Hz								
	63	125	250	500	1000	2000	4000	8000
L_W, dB	91.0	87.0	83.0	82.0	78.0	76.0	72.0	69.0
Dist. Corr., dB	−24.2	−24.2	−24.2	−24.2	−24.2	−24.2	−23.5	−23.5
ΔA, dB	−7.5	−9.1	−11.1	−13.3	−15.8	−18.4	−21.1	−23.8
ΔL, dB	1.9	1.9	1.9	1.9	1.9	1.9	1.9	1.9
DI, dB	3.0	3.0	3.0	3.0	3.0	3.0	3.0	3.0
L_p @ Rec., dB	64.2	58.6	52.6	49.4	42.9	38.3	31.6	25.9

GENERAL INFORMATION ON THE DESIGN OF HVAC

CHAPTER 8

8.1 GENERAL DESIGN GUIDELINES

Background sound from an HVAC system in a building should be unobtrusive and have low enough levels so that it does not interfere with the requirements of the spaces being served. The sound should be bland and have no particular identity with frequency. There should be a balanced distribution of sound energy over a broad frequency range. In general the 1/1 octave band sound levels should decrease at a rate of 5 dB per 1/1 octave frequency band. This will result in a spectrum similar to the RC curves shown in Figure 2.3. There should be no audible tonal characteristics, such as hum, whistle, whine, or rumble, associated with the spectrum. Finally, there should be no detectable time variations in the levels of the spectrum or parts of the spectrum that are associated with beats or other system-induced aerodynamic instabilities.

With respect to the quality of sound associated with HVAC system noise in an occupied space, fan noise generally contributes to the sound levels in the 63 Hz through 250 Hz 1/1 octave frequency bands. This is shown in Figure 8.1 as curve A. Diffuser noise usually contributes to the overall HVAC noise in the 250 Hz through 8,000 Hz 1/1 octave frequency bands. This is shown as curve B in Figure 8.1. The overall sound pressure levels associated with both the fan and diffuser noise are shown as curve D. The RC level of the overall noise is RC 36. The RC 36 curve is superimposed over curve D. As can be seen by comparing the RC curve with curve D, the classification of the overall noise is neutral. Curve D represents what would be considered acceptable and desirable 1/1 octave band sound pressure levels in many occupied spaces.

Several general factors should be considered when selecting fans and other related equipment and when designing air distribution systems to minimize the noise transmitted from different components of the system to the occupied spaces it serves. They include:

1. Air distribution systems should be designed to minimize flow resistance and turbulence. High flow resistance increases the required fan pressure, which results in higher noise being generated by the fan. Turbulence increases the flow noise generated by duct fittings and dampers in the air distribution system.
2. A fan should be selected to operate as near as possible to its rated peak efficiency when handling the required quantity of air and static pressure. Also, a fan should be selected which generates the lowest possible noise but still meets the required design conditions for which it is selected. Oversized or undersized fans that do not operate at or near rated peak efficiencies result in substantially higher noise levels.
3. Duct connections at both the fan inlet and outlet should be designed for uniform and straight air flow. Failure to do this can result in severe turbulence at the fan inlet and outlet and in flow separation at the fan blades. Both of these can significantly increase the noise generated by the fan.
4. Care should be exercised when selecting duct silencers to attenuate supply or return air noise. Duct silencers can significantly increase the required fan static pressure. When a rectangular duct silencer is used, it is necessary to line the duct for a distance of at least 10 ft beyond the silencer with a minimum 1-in.-thick fiberglass duct lining to reduce high-frequency regenerated noise associated with the silencer. Whenever possible, acoustically lined sound plenums should be used in the place of duct silencers.
5. Fan-powered mixing boxes associated with variable-volume air distribution systems should not be placed over or near noise-sensitive areas.
6. Air flowing by or through elbows or duct branch take-offs generate turbulence. To minimize the flow noise associated with this turbulence, elbows and duct branch take-offs should be located at least four to five duct diameters from each other whenever possible. For high velocity systems, it may be necessary to increase this distance to up to 10 duct diameters in critical noise areas.
7. Near critical noise areas, it may be desirable to expand the duct cross-sectional area to keep the airflow velocity in the duct as low as possible. This will reduce potential flow noise associated with turbulence in these areas.

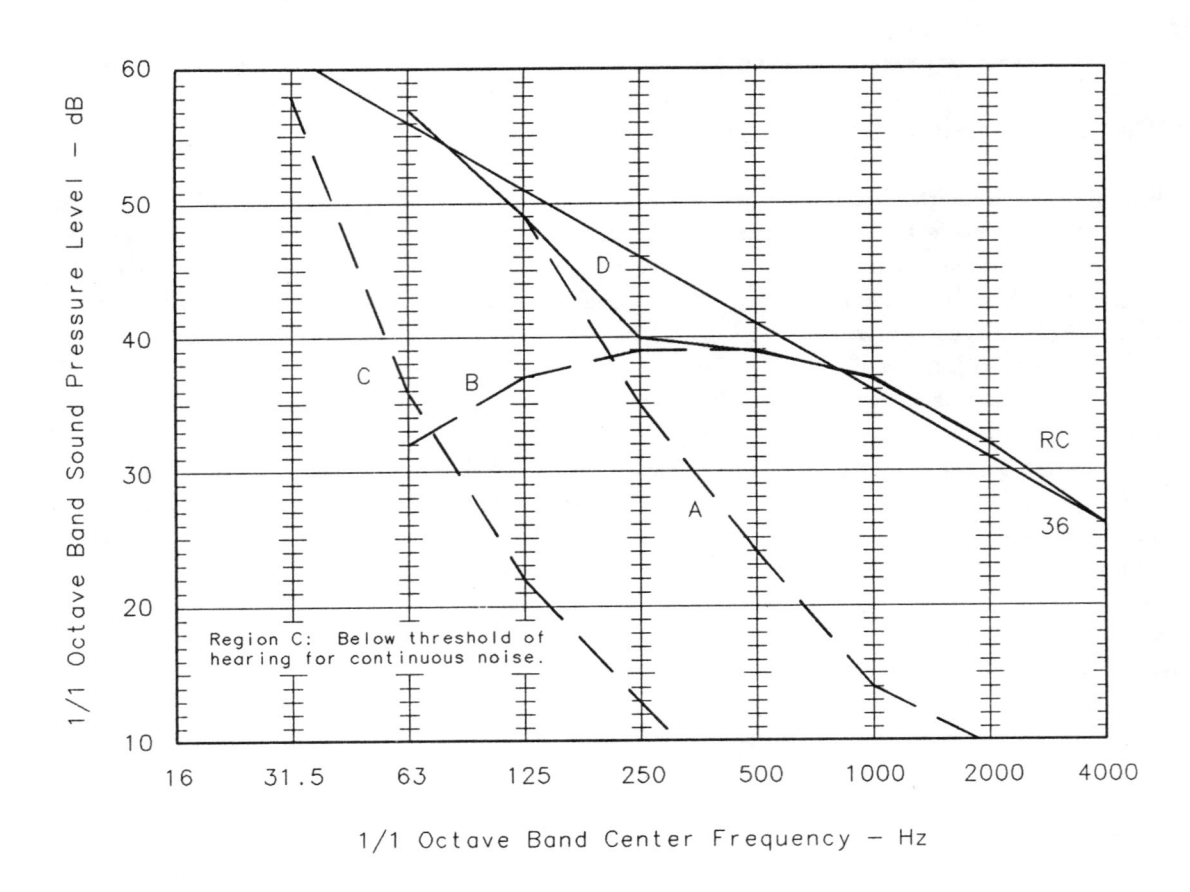

Figure 8.1 Illustration of Well–Balanced HVAC Sound Spectrum for Occupied Spaces

8. Turning vanes should be used in large, 90° rectangular elbows and branch take–offs. This provides a smoother transition in which the air can change flow direction, thus reducing turbulence.
9. Grilles, diffusers, and registers into occupied spaces should be placed as far as possible from elbows and branch take–offs.
10. Grilles, diffusers and registers should not be placed near room corners or edges where wall and ceiling or wall and floor meet.

Table 8.1 lists several common sound sources associated with mechanical equipment noise. Anticipated sound transmission paths and recommended noise reduction methods are also listed in the table. Airborne and/or structure–borne noise can follow any or all of the transmission paths associated with a specified sound source.

In order to effectively deal with each of the different sound sources and related sound paths associated with an HVAC system, the following design procedures are suggested:

1. Determine the design goal for HVAC system noise for each critical area according to its use and construction. Use Table 2.1 to specify the desirable NC or RC levels.
2. Relative to equipment that radiates sound directly into a room, select equipment that will be quiet enough to meet the desired design goal.
3. If central or roof–mounted mechanical equipment is used, complete an initial design and layout of the HVAC system using acoustical treatment where it appears appropriate.
4. Starting at the fan, appropriately add the sound attenuations and sound power levels associated with the central fan(s), fan–powered mixing units (if used), and duct elements between the central fan(s) and the room of interest to determine the corresponding sound pressure levels in the room. Be sure to investigate the

GENERAL INFORMATION ON THE DESIGN OF HVAC SYSTEMS

Table 8.1 Sound Sources, Transmission Paths, and Recommended Noise Reduction Methods

Sound Source	Path Nos.
Circulating fans; grilles; diffusers; registers; unitary equipment in room	1
Induction coil and fan–powered mixing units	1,2
Unitary equipment located outside of room served; remotely located air–handling equipment such as fans and blowers, dampers, duct fittings, and air washers	2,3
Compressors and pumps	4,5,6
Cooling towers; air–cooled condensers	4,5,6,7
Exhaust fans; window air conditioners	7,8
Sound transmission between rooms	9,10

Transmission Paths		Recommended Noise–Reduction Methods
No.	Description	
1	Direct sound radiated from sound source to ear. Reflected sound from walls, ceiling, and walls	Direct sound can be controlled only by selecting quiet equipment. Reflected sound is controlled by adding sound absorption to room and to location of equipment
2	Air and structure borne sound radiated from casings and through walls of ducts and plenums is transmitted through walls and ceiling into room	Design ducts and fittings for low turbulence; locate high velocity ducts in noncritical areas; isolate ducts and sound plenums from structure with neoprene or spring hangers
3	Airborne sound radiated through supply and return air ducts to diffusers in room and then to listener by Path 1	Select fans for minimum sound power; use ducts lined with sound–absorbing material; use duct silencers or sound plenums in supply and return air ducts
4	Noise is transmitted through equipment room walls and floors to adjacent rooms	Locate equipment rooms away from critical areas; use masonry blocks or concrete for equipment room walls and floor
5	Building structure transmits vibration to adjacent walls and ceilings, from which it is radiated as noise into room by Path 1	Mount all machines on properly designed vibration isolators; design mechanical equipment room for dynamic loads; balance rotating and reciprocating equipment
6	Vibration transmission along pipe and duct walls	Isolate pipe and ducts from structure with neoprene or spring hangers; install flexible connectors between pipes, ducts, and vibrating machines
7	Noise radiated to outside enters room windows	Locate equipment away from critical areas; use barriers and covers to interrupt noise paths; select quiet equipment
8	Inside noise follows Path 1	Select quiet equipment
9	Noise transmitted to diffuser in a room into ducts and out through an air diffuser in another room	Design and install duct attenuation to match transmission loss of wall between rooms
10	Sound transmission through, over, and around room partitions	Extend partition to ceiling slab and tightly seal all around; seal all pipe, conduit, and duct penetrations

supply and return air paths. Investigate possible duct sound breakout when central fans are adjacent to the room of interest or roof—mounted fans are above the room of interest.

5. If the mechanical equipment room is adjacent to the room of interest, determine the sound pressure levels in the room associated with sound transmitted through the mechanical equipment room wall.
6. Add the sound pressure levels in the room of interest that are associated with all of the sound paths between the mechanical equipment room or roof—mounted unit and the room of interest.
7. Determine the corresponding NC or RC level associated with the calculated total sound pressure levels in the room of interest.
8. If the NC or RC level exceeds the design goal, determine the 1/1 octave frequency bands in which the corresponding sound pressure levels are exceeded and the sound paths that are associated with these 1/1 octave frequency bands.
9. Redesign the system, adding additional sound attenuation to the paths that contribute to the excessive sound pressure levels in the room of interest.
10. Repeat Steps 4 through 9 until the desired design goal is achieved.
11. Steps 3 through 10 must be repeated for every room that is to be analyzed.
12. Make sure that noise radiated by outdoor equipment will not disturb adjacent properties.
13. With respect to outdoor equipment, use barriers when noise associated with the equipment will disturb adjacent properties if barriers are not used.
14. If mechanical equipment is located on upper floors or is roof—mounted, vibration isolate all reciprocating and rotating equipment. It may be necessary to vibration isolate mechanical equipment that is located in the basement of a building.
15. If possible, use flexible connectors between rotating and reciprocating equipment and pipes and ducts that are connected to the equipment.
16. If it is not possible to use flexible connectors between rotating and reciprocating equipment and pipes and ducts connected to the equipment, use spring or neoprene hangers to vibration isolate the ducts and pipes within the first 20 ft of the equipment.
17. Use either flexible connectors or spring or neoprene hangers. Do not use both.
18. Use flexible conduit between rigid electrical conduit and reciprocating and rotating equipment.

8.2 SYSTEM EXAMPLE

Individual examples have been given in the preceding sections that demonstrate how to calculate equipment and regenerated sound power levels and sound attenuation values associated with the system elements of HVAC air distribution systems. It is now worth while to examine a complete HVAC system example to see how the information that has been presented can be tied together to determine the sound pressure levels associated with a specific HVAC system. Complete calculations for each system element will not be given. Only a summary of the tabulated results will be listed.

Air is supplied to the HVAC system in this example by the rooftop unit shown in Figure 8.2. The receiver room is a room that is directly below the unit. The room has the following dimensions: length — 20 ft, width — 20 ft; and height — 9 ft. For this example, it is assumed that the roof penetrations associated with the supply and return air ducts are well sealed and that there are no other roof penetrations associated with the unit. The supply side of the rooftop unit is ducted to a VAV control unit that serves the room in question. A return air grill conducts air to a common ceiling return air plenum. The return air is then directed to the rooftop unit through a short rectangular return air duct.

Three sound paths are examined:

Path 1.   Fan airborne supply air sound that enters the room through the ceiling diffuser.
Path 2.   Fan airborne supply air sound that breaks out through the duct wall of the main supply air duct in the plenum space above the room.
Path 3.   Fan airborne return air sound that enters the room from the inlet of the return air duct.

GENERAL INFORMATION ON THE DESIGN OF HVAC SYSTEMS

8–5

The sound power levels associated with the supply air and return air sides of the fan in the rooftop unit are specified by the manufacturer to be:

	\multicolumn{7}{c}{1/1 Octave Band Center Freq – Hz}
	63	125	250	500	1K	2K	4K
ROOFTOP SUPPLY AIR; CFM 7000; SP 2.5IN	92	86	80	78	78	74	71
ROOFTOP RETURN AIR; CFM 7000; SP 2.5IN	82	79	73	69	69	67	59

Paths 1 and 2 are associated with the supply–air side of the system. Figure 8.3 shows a layout of the part of the supply air system that is associated with the receiver room. The main duct is a 22 in.–diameter, 26 gauge, unlined, circular sheet metal duct. The flow volume in the main duct is 7,000 cfm. The silencer after the radiused elbow is a 22–in.–diameter by 44 in. long, high–pressure, circular silencer. The branch junction that occurs 8 feet from the silencer is a 45° wye. The branch duct between the main duct and the VAV control unit is a 10–in.–diameter, unlined, circular sheet metal duct. The flow volume in the branch duct is 800 cfm. The straight section of duct between the VAV control unit and the diffuser is a 10–in.–diameter, unlined circular sheet metal duct. The diffuser is a 15 in. by 15 in. square diffuser. Assume a typical distance between the diffuser and a listener in the room is 5 ft. With regard to the duct breakout sound associated with the main duct, the length of the duct that runs over the room is 20 ft. The ceiling of the room is comprised of 2 ft by 4 ft by 5/8 in. lay–in ceiling tiles that have a surface weight of 0.6–0.7 lb/ft^2. The ceiling has integrated lighting and diffusers. Path 3 is associated with the return air side of the system. Figure 8.4 shows a layout of the part of the return air system that is associated with the receiver room. The rectangular return air duct is lined with 2 in.–thick 3 lb/ft^3 density fiberglass duct liner. For the return air path, assume the typical distance between the inlet of the return air duct and a listener is 10 ft.

The analysis associated with each path begins at the rooftop unit (fan) and progressively proceeds through the different system elements to the receiver room. The system element numbers in the tables correspond to the element numbers contained in brackets in Figures 8.3 and 8.4.

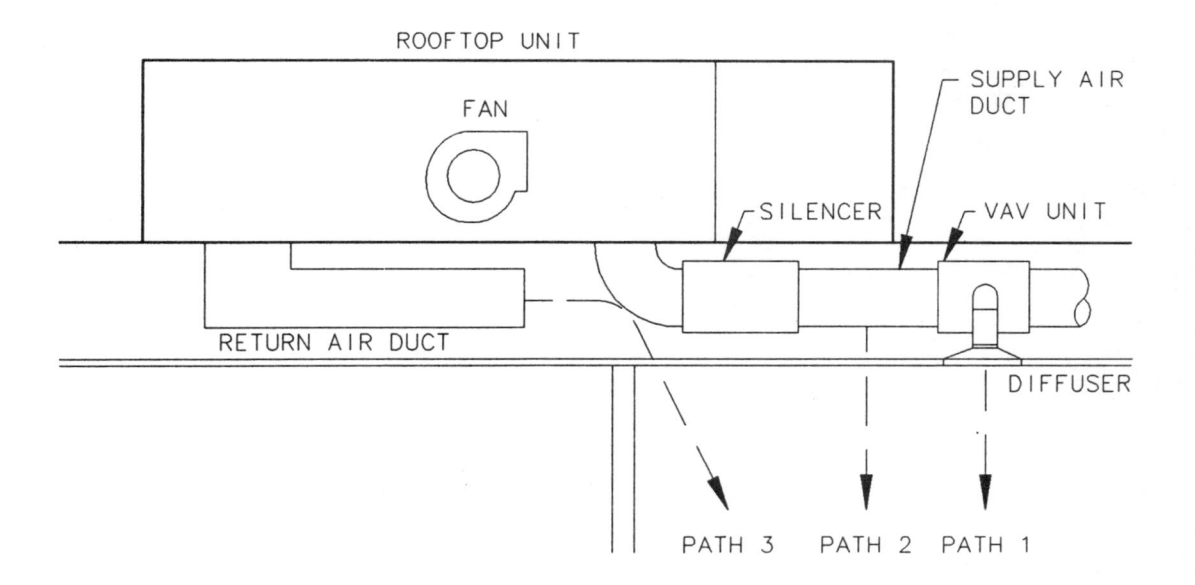

Figure 8.2 Paths for System Example

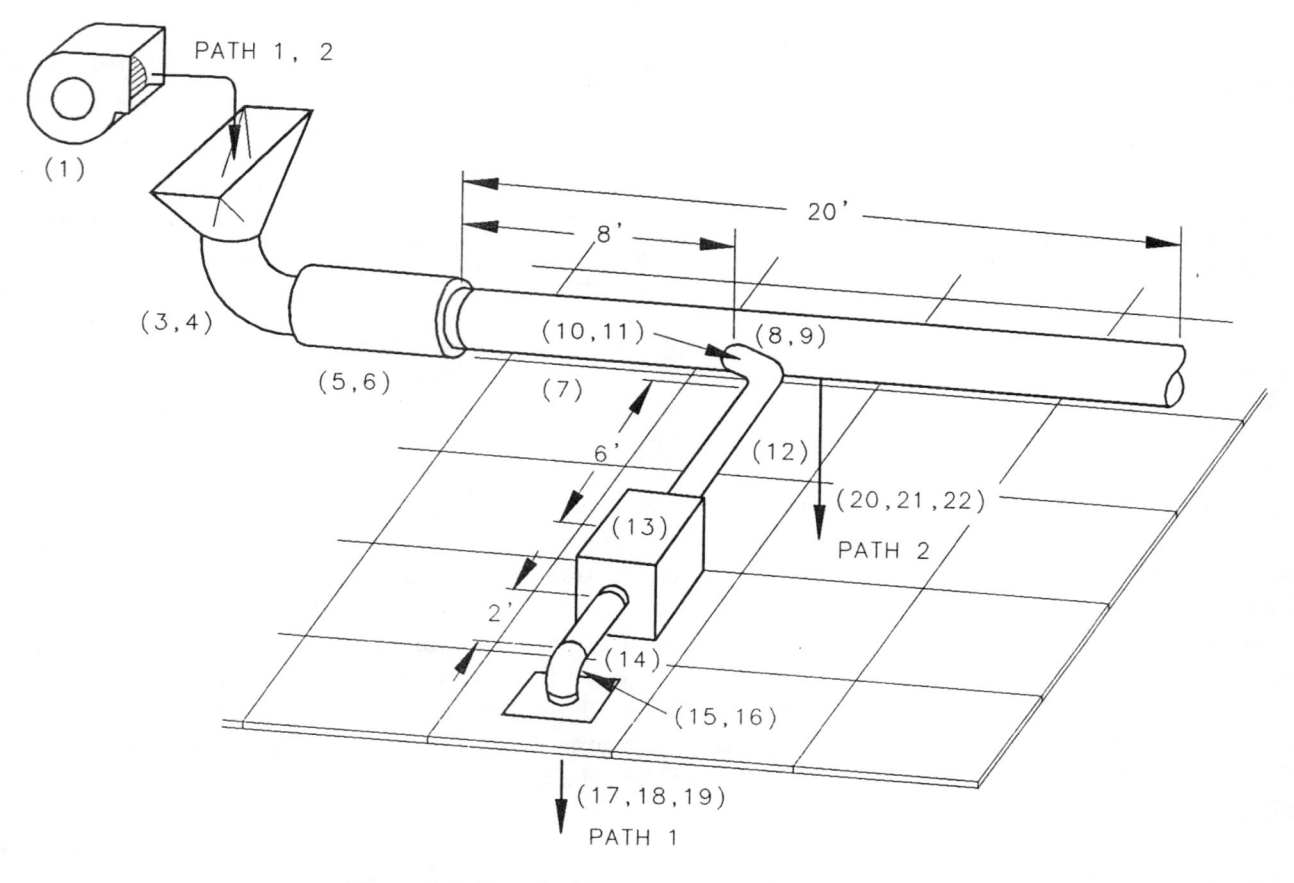

8-6 GENERAL INFORMATION ON THE DESIGN OF HVAC SYSTEMS

Figure 8.3 Supply Air Portion of System Example

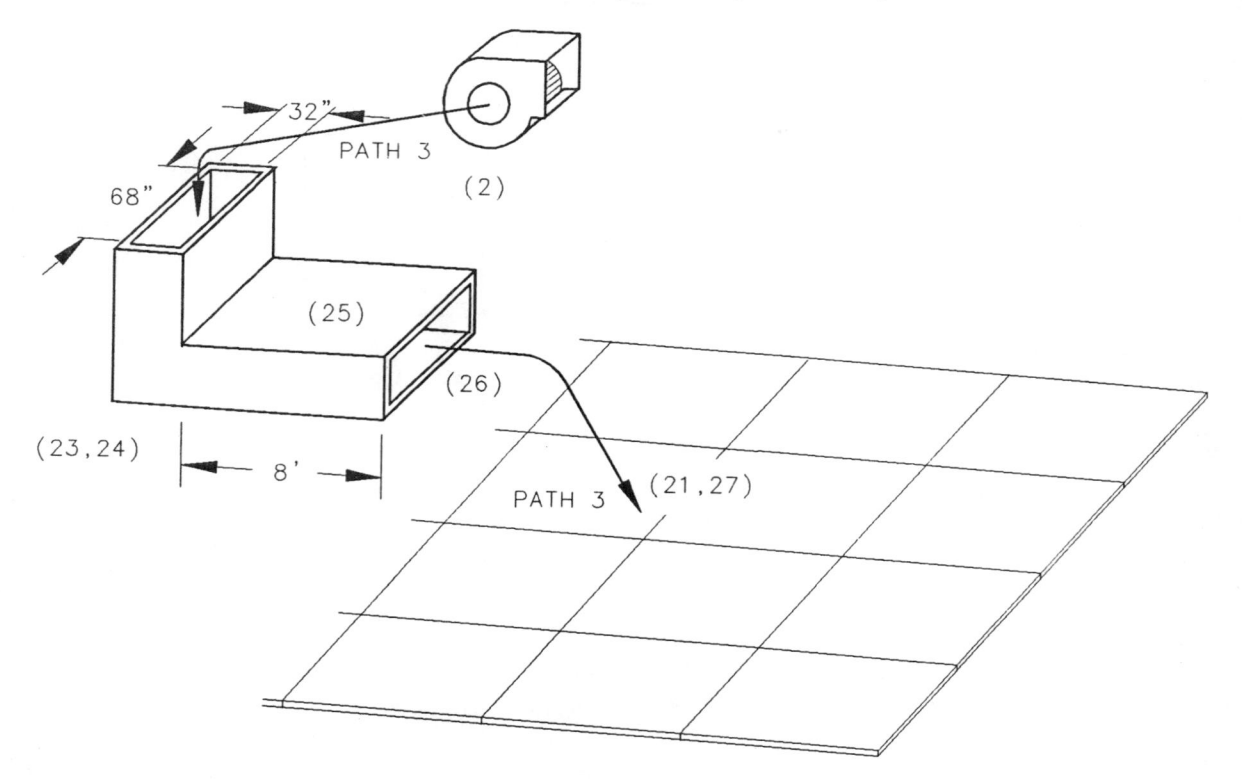

Figure 8.4 Return Air Portion of System Example

GENERAL INFORMATION ON THE DESIGN OF HVAC SYSTEMS

PATH 1

DESCRIPTION	| 1/1 Octave Band Center Freq – Hz						
	63	125	250	500	1K	2K	4K
1 FAN – SUPPLY AIR; CFM 7000; SP 2.5IN	92	86	80	78	78	74	71
3 22IN WIDE(DIA) UNLINED RADIUS ELBOW	0	–1	–2	–3	–3	–3	–3
SUM WITH NOISE REDUCTION VALUES	92	85	78	75	75	71	68
4 90 DEG BEND W/O TURNING VANES; 12IN RAD	56	54	51	47	42	37	29
SUM SOUND POWER LEVELS	92	85	78	75	75	71	68
5 22 IN DIA X 44 IN HIGH PRESS SILENCER	–4	–7	–19	–31	–38	–38	–27
SUM WITH NOISE REDUCTION VALUES	88	78	59	44	37	33	41
6 REG NOISE FROM ABOVE SILENCER	68	79	69	60	59	59	55
SUM SOUND POWER LEVELS	88	82	69	60	59	59	55
7 22IN DIA X 8FT UNLINED CIR DUCT	0	0	0	0	0	0	0
10 BR PWL DIV; M–22IN DIA; B–10IN DIA	–8	–8	–8	–8	–8	–8	–8
SUM WITH NOISE REDUCTION VALUES	80	74	61	52	51	51	47
1 DUCT 90 DEG BRANCH TAKEOFF; 2IN RADIUS	56	53	50	47	43	37	31
SUM SOUND POWER LEVELS	80	74	61	53	52	51	47
12 10IN DIA X 6FT UNLINED CIR DUCT	0	0	0	0	0	0	0
13 TERMINAL VOLUME REG UNIT (GEN ATTN)	0	–5	–10	–15	–15	–15	–15
14 10IN DIA X 2FT UNLINED CIR DUCT	0	0	0	0	0	0	0
15 10IN WIDE(DIA) UNLINED RADIUS ELBOW	0	0	–1	–2	–3	–3	–3
SUM WITH NOISE REDUCTION VALUES	80	69	50	36	34	33	29
16 90 DEG BEND W/O TURNING VANES; 2IN RAD	49	45	41	37	31	24	16
SUM SOUND POWER LEVELS	80	69	51	40	36	34	29
17 10IN DIA DIFFUSER END REF LOSS	–16	–10	–6	–2	–1	0	0
SUM WITH NOISE REDUCTION VALUES	64	59	45	38	35	34	29
18 15IN X 15IN RECTANGULAR DIFFUSER	31	36	39	40	39	36	30
SUM SOUND POWER LEVELS	64	59	46	42	40	38	33
19 ASHRAE ROOM CORR – 1 IND SOUND SOURCE	–5	–6	–7	–8	–9	–10	–11
SOUND PRESS LEVELS – RECEIVER ROOM	59	53	39	34	31	28	22

NC: NC = 36   RC: RC = 31(R)   DBA: 41 DBA

The first table is associated with Path 1. The first entry in the table is the manufacturer's values for the supply air fan sound power levels (1). The second entry is the sound attenuation associated with the 22–in.–diameter unlined radius elbow (3). Since the next entry is associated with the regenerated sound power levels associated with the elbow (4), it is necessary to tabulate the results associated with the elbow attenuation to determine the sound power levels at the exit of the elbow. These sound power levels and the elbow regenerated sound power levels are then added logarithmically. In a like fashion, the dynamic insertion loss values of the duct silencer (5) and the silencer regenerated sound power levels (6) are included in the table and tabulated. Next, the attenuation associated with the 8–ft section of 22–in.–diameter duct (7) and the branch power division (10) associated with sound propagation in the 10–in.–diameter branch duct are included in the table. After element 10, the sound power levels that exist in the branch duct after the branch takeoff are calculated so that the regenerated sound power levels (11) in the branch duct associated with the branch takeoff can be logarithmically added to the results. Next, the sound attenuation values associated with the 6–ft section of 10–in.–diameter unlined duct (12), the terminal volume regulation unit (13), the 2 foot section of 10–in.–diameter unlined duct (14), and 10–in.–diameter radius elbow (15) are included in the table. The sound power levels that exist at the exit of the elbow are then calculated so that the regenerated sound power levels (16) associated with the elbow can be logarithmically added to the results. The diffuser end reflection loss (17) and the diffuser regenerated sound power levels (18) are appropriately included in the table. The sound power levels that are tabulated after element 18 are the sound power levels that exist at the diffuser in the receiver room. The final entry in the table is the "room correction," which converts the sound power levels at the diffuser to their corresponding sound pressure levels at the point of interest in the receiver room. The NC, RC, and dBA values associated with the sound pressure levels from Path 1 are listed as the last line in the table.

PATH 2

DESCRIPTION	\multicolumn{7}{c}{1/1 Octave Band Center Freq – Hz}
	63	125	250	500	1K	2K	4K
1 FAN – SUPPLY AIR; CFM 7000; SP 2.5IN	92	86	80	78	78	74	71
3 22IN WIDE(DIA) UNLINED RADIUS ELBOW	0	−1	−2	−3	−3	−3	−3
SUM WITH NOISE REDUCTION VALUES	92	85	78	75	75	71	68
4 90 DEG BEND W/O TURNING VANES; 12IN RAD	56	54	51	47	42	37	29
SUM SOUND POWER LEVELS	92	85	78	75	75	71	68
5 22 IN DIA X 44 IN HIGH PRESS SILENCER	−4	−7	−19	−31	−38	−38	−27
SUM WITH NOISE REDUCTION VALUES	88	78	59	44	37	33	41
6 REG NOISE FROM ABOVE SILENCER	68	79	69	60	59	59	55
SUM SOUND POWER LEVELS	88	82	69	60	59	59	55
7 22IN DIA X 8FT UNLINED CIR DUCT	0	0	0	0	0	0	0
8 BR PWL DIV; M−22 IN DIA; B−22 IN DIA	−1	−1	−1	−1	−1	−1	−1
SUM WITH NOISE REDUCTION VALUES	87	81	68	59	58	58	54
9 DUCT 90 DEG BRANCH TAKEOFF; 2IN RADIUS	63	60	57	54	50	44	34
SUM SOUND POWER LEVELS	87	81	68	60	59	58	54
20 22IN DIA X 20FT 26 GA DUCT BREAKOUT	−29	−29	−21	−11	−9	−7	−5
21 2X4X5/8 LAY-IN CEILING	−4	−8	−8	−12	−14	−15	−15
22 LINE SOURCE—MED DEAD ROOM	−6	−5	−4	−6	−7	−8	−9
SOUND PRESS LEVELS – RECEIVER ROOM	48	39	35	31	29	28	25

NC: NC = 29   RC: RC = 29(H)   DBA: 35 DBA

Elements 1 through 7 in Path 2 are the same as Path 1. Elements 8 and 9 are associated with the branch power division (8) and the corresponding regenerated sound power levels (9) of the sound that propagates down the main duct beyond the duct branch. The next three entries in the table are the sound transmission loss associated with the duct breakout sound (20), the sound transmission loss associated with the ceiling (21), and the "room correction" (22) that converts the sound power levels at the ceiling to corresponding sound pressure levels in the room.

PATH 3

DESCRIPTION	\multicolumn{7}{c}{1/1 Octave Band Center Freq – Hz}
	63	125	250	500	1K	2K	4K
1 FAN – RETURN AIR; CFM 7000; SP 2.5IN	82	79	80	78	78	74	71
23 32IN WIDE LINED SQ ELBOW W/O TURN VANES	−1	−6	−11	−10	−10	−10	−10
SUM WITH NOISE REDUCTION VALUES	81	73	69	68	68	64	61
24 90 DEG BEND W/O TURNING VANES; .5IN RAD	77	73	68	62	55	48	38
SUM SOUND POWER LEVELS	82	76	72	69	68	64	61
25 32IN X 68IN X 8FT LINED DUCT	−2	−2	−6	−18	−15	−12	−13
26 32IN X 68IN DIFFUSER END REF LOSS	−4	−2	0	0	0	0	0
21 2X4X5/8 LAY-IN CEILING	−4	−8	−8	−12	−14	−15	−15
27 ASHRAE ROOM CORR −1 IND SOUND SOURCE	−8	−9	−10	−11	−12	−13	−14
SOUND PRESS LEVELS – RECEIVER ROOM	64	55	48	28	27	24	19

NC: NC = 40   RC: RC = 26(R)   DBA: 44 DBA

The first element in Path 3 is the manufacturer's values for return air fan sound power levels (2). The next two elements are the sound attenuation associated with a 32−in.−wide lined square elbow without turning vanes (23) and the regenerated sound power levels associated with the square elbow (24). The final four elements are the insertion loss associated with a 32 in. by 68 in. by 8 ft rectangular sheet metal duct lined with 2−in.−thick 3 lb/ft^3

fiberglass duct lining (26), the diffuser end reflection loss (27), the transmission loss through the ceiling (21), and the "room correction" (27) that converts the sound power levels at the ceiling to corresponding sound pressure levels in the room.

The total sound pressure levels in the receiver room from the three paths are obtained by logarithmically adding the individual sound pressure levels associated with each path. The NC value in the room associated with sound from all three paths is NC 40; the RC value is RC 33(R–H); and the A weighted sound pressure level is 44 dBA.

TOTAL SOUND PRESSURE LEVELS – ALL PATHS

DESCRIPTION	63	125	250	500	1K	2K	4K
SOUND PRESSURE LEVELS PATH NO. 1	59	53	39	34	31	28	22
SOUND PRESSURE LEVELS PATH NO. 2	48	39	35	31	29	28	25
SOUND PRESSURE LEVELS PATH NO. 3	65	55	48	28	27	24	19
TOTAL SOUND PRESSURE LEVELS – ALL PATHS	65	57	49	37	34	32	28

1/1 Octave Band Center Freq – Hz

NC: NC = 40   RC: RC = 26(R)   DBA: 44 DBA

CHAPTER 9
CONCLUSIONS AND RECOMMENDATIONS

This project has pulled together nearly all relevant information related to HVAC acoustics that currently exists in the open literature. The following are conclusions relative to this information.

FANS

The information on fan sound power levels that is presented in Section 3.1 is from the material on fans included in the last two revisions of the sound and vibration control chapter of the ASHRAE Systems Handbook (36,38). A comparison of this information with selected manufacturers' fan data has indicated that there are areas of significant disagreement between the ASHRAE fan data and corresponding manufacturers' data. Thus, the ASHRAE fan data should be adjusted to eliminate, as much as possible, these areas of disagreement.

DUCT ELEMENT REGENERATED SOUND POWER

Information pertaining to duct element regenerated sound power associated with dampers, elbows fitted with turning vanes, and junctions and turns appears to be good. Work reported by Ver and contained in the sound and vibration control chapter of the 1987 ASHRAE HVAC Systems and Applications Handbook is reasonably thorough in this area (36,44,45,46,47). Typically, HVAC duct systems that are properly designed do not have significant duct element regenerated noise problems associated with duct fittings. Thus, there appears to be no real need for additional work in this area.

Information relative to sound power generated by diffusers is not very complete. Manufacturers' product data usually only list the noise levels associated with diffusers in the form of NC levels. There are little or no published experimental data relative to the sound power levels associated with diffusers. In well-designed HVAC systems, the high-frequency noise radiated from the system into a room is usually associated with noise from the supply or return air diffusers. Thus, this information is important. The procedures presented in this publication generally yield NC levels that correlate reasonably well with manufacturers' published NC levels for generic-type diffusers only. They do not yield results that correlate well with diffusers that have specially designed plenum and/or damper systems. Also, even though there may be good correlation with respect to NC levels obtained from the procedures described in this publication and corresponding published manufacturers' data, there are no published data to indicate whether or not the same correlation exists relative to the corresponding 1/1 octave band sound power levels. Work should be initiated to yield more complete information in this area.

DUCT ELEMENT SOUND ATTENUATION

No new information has been presented on sound plenums since Wells published his paper in 1958 (48). Sound plenums are often used to attenuate low-frequency sound from fans. There currently are no good data relative to the low frequency attenuation characteristics of plenums. Also, all of the information that is available is for the case where the plenum outlet is parallel to the plenum inlet and for the case where there is only one plenum outlet. There are many situations where the plenum outlet is perpendicular to the the plenum inlet and where there is more than one plenum outlet for an individual plenum. Since sound plenums are often an important element in controlling low-frequency sound in HVAC systems, work should be initiated to generate accurate information in these areas.

Experimental data for unlined rectangular ducts are old, irregular, and difficult to correlate (10, 14,29,39,45). The information that is available indicates that rectangular ducts can have a significant impact on the attenuation of low-frequency sound in HVAC systems. However, there is some question relative to the accuracy of these data. Also, the results of using insertion loss data for acoustically lined rectangular ducts indicate that unlined rectangular ducts can have higher sound attenuation at low frequencies than correspondingly sized acoustically lined rectangular ducts. There currently is no reliable information on how the sound attenuation associated with unlined rectangular ducts can be properly combined with the measured insertion loss data for acoustically lined rectangular ducts to give the total sound attenuation of the ducts. Work should be initiated to generate this information.

CONCLUSIONS AND RECOMMENDATIONS

Experimental data for acoustically lined rectangular and circular ducts are very recent and appear to be good. The data are for no flow conditions (8,22,23,24). Work should be initiated to determine the effects of airflow on sound attenuation in acoustically lined rectangular and circular ducts.

Information associated with the sound attenuation of rectangular or radiused, unlined sheet metal elbows, and for acoustically lined, rectangular sheet metal elbows is sparse and very old. The information that is available is based on Figure 5.2, which is taken from Beranek's book, Noise Reduction, published in 1960 (4). Work should be initiated to obtain current and accurate information relative to the sound attenuation provided by rectangular and radiused, unlined sheet metal elbows and acoustically lined, rectangular sheet metal elbows.

Recent data were used to determine the equations used to calculate the sound attenuation of acoustically lined, radiused, circular elbows (8). These data appear to be good and reasonably accurate. The data that were analyzed were obtained under no–flow conditions. Data are available and should be analyzed to determine the effects of air flow on the sound attenuation associated with acoustically lined, radiused, circular elbows.

Product data for duct silencers are quite substantial. Although data for duct silencers may be a little optimistic at times, these are the best data available and should be used when determining the sound attenuation through duct silencers.

Information for duct branch sound power attenuation appears to be suitable. Work reported by Ver appears to cover branch sound power division reasonably well (44,46).

Very few experimental data relative to the low–frequency sound attenuation associated with duct–end reflection losses are available in the open literature. Work reported by Sandbakken et al., which was the basis for AMCA Standard 300–85, was used to describe the effects of acoustical duct–end reflection losses (1,32). The data used applied only to round ducts or low aspect ratio rectangular ducts and to open ducts that either terminate flush with a wall or terminate in free space. The data did not take into account the effects of diffusers placed over the end of the ducts. Since duct–end reflection loss accounts for a significant amount of the low–frequency sound attenuation in HVAC duct systems, work should be initiated to determine the effects of high–aspect–ratio rectangular ducts and of diffusers on acoustical duct end reflection losses.

DUCT BREAKOUT AND BREAKIN

Results reported by Cummings for duct breakout and breakin in rectangular, circular, and flat oval ducts were made under no–flow conditions (11,12). There is significant justification to indicate that airflow–induced, low–frequency vibration of duct walls is a significant source of low–frequency duct rumble and breakout sound. The results reported by Cummings did not take this into account. The actual transmission loss values in areas where airflow–induced duct vibration is present can be significantly lower than those values reported by Cummings. This should be investigated.

Work performed by Cummings on the insertion loss of externally lagged rectangular ducts tends to give optimistic results. Experimental work reported by Harold relative to the insertion loss of externally lagged rectangular ducts indicated that the values he obtained were less than the corresponding values predicted by Cummings (19). Work should be performed in this area to clear up these discrepancies.

SOUND TRANSMISSION IN INDOOR SPACES

Single–pass data for acoustical ceilings are very limited, particularly for integrated ceiling systems. Work should to be performed to gather more data on integrated acoustical ceiling systems.

Significant questions with regard to low–frequency acoustical properties of rooms exist. There are some discrepancies that exist between the results of experimental studies reported by Schultz and Thompson on similar spaces (33,42.43). Questions also exist relative to the low–frequency calibration of reference sound sources similar to the one used to generate the data reported by Schultz. There currently are no data to clearly justify extrapolating Shultz's or Thompson's data down to a 1/1 octave band center frequency of 63 Hz. Work should be initiated to clear up the questions concerning the low–frequency calibration of reference sound sources, to clear up the discrepancies in the results reported by Schultz and Thompson, and to yield information that can be used to describe the acoustical properties of rooms down to a 1/1 octave band center frequency of 63 Hz and possibly to 31.5 Hz.

REFERENCES

1. Beatty, J. 1987. "Discharge duct configurations to control rooftop sound." Heating/Piping/Air Conditioning July.

2. Beranek, L. L. 1954. Acoustics, New York: McGraw–Hill.

3. Beranek, L. L. 1971. Noise and Vibration Control, New York: McGraw–Hill.

4. Beranek, L. L. 1960. Noise Reduction, New York: McGraw–Hill.

5. Beranek, L. L. 1957. "Revised criteria for noise in buildings," Noise Control, January.

6. "Big–H, quietflow plenums." 1980. Industrial Acoustics Company, Bulletin 1.0207.3.

7. Blazier, W. E., Jr. 1981. "Revised noise criteria for design and rating of HVAC systems," ASHRAE Transactions, Vol. 87, Part 1.

8. Bodley, J. 1988. personal communication, The United McGill Corporation, Groveport, OH.

9. Catalog of STC and IIC ratings for wall and floor/ceiling assemblies. 1981. Office of Noise Control, California Department of Health Services.

10. Chaddock, J. B. et al. 1959. "Sound attenuation in straight ventilation ducting," Refrigeration Engineering, January.

11. Cummings, A. 1983. "Acoustic noise transmission through the walls of air conditioning ducts," Final Contract Report, Department of Mechanical and Aerospace Engineering, University of Missouri–Rolla, December.

12. Cummings, A. 1985. "Acoustic noise transmission through duct walls," ASHRAE Transactions, Vol. 91, Part 2A.

13. Cummings, A. 1979. "The effects of external lagging on low frequency sound transmission through the walls of rectangular ducts," Journal of Sound and Vibration, Vol. 67, No. 2, pp. 187–201.

14. "Design for sound." 1973. Woods Fan Division, The English Electric Corporation.

15. Diehl, G. M. 1973. Machinery acoustics. New York: John Wiley & Sons.

16. Faulkner, L. L., ed. 1976. Handbook of industrial noise control. pp. 419–424. Industrial Press, Inc.

17. Galloway, W. J. and Schultz, T. J. 1979. "Noise assesment guideliness – 1979 technical background." Bolt, Beranek and Newman, Inc., Report 4024, August.

18. Goodfriend, L. S. 1980. "Indoor sound rating criteria." ASHRAE Transactions, Vol. 86, Part 2.

19. Harold, R. G. 1986. "Round duct can stop rumble noise in air–handling installations." ASHRAE Transactions, Vol. 92, Part 2.

20. Harris, C. M., ed. 1971. Handbook of noise control. New York: McGraw–Hill.

21. Hirschorn, M. 1981. "Acoustic and aerodynamic characteristics of duct silencers for air–handling installations." ASHRAE Transactions, Vol. 87, Part 1.

22. Kuntz, H. L. 1986. "The determination of the interrelationship between the physical and acoustical properties of fibrous duct liner materials and lined duct sound attenuation." Report No. 1068, Hoover Keith & Bruce, Inc., December.

23. Kuntz, H. L., and R. M. Hoover. 1987. "The interrelationships between the physical properties of fibrous duct lining materials and lined duct sound attenuation." ASHRAE Transactions, Vol. 93, Part 2.

24. Machen, J., and J. C. Haines. 1983. "Sound insertion loss properties of linacoustic and competitive duct liners." Report No. 436–T–1778, Johns–Manville ReseaDevelopment Center, December.

25. Owens–Corning Fiberglass Corp. 1981. "Noise control manual." , 4th Edition.

26. Pallet, D. S., W. Robert, R. D. Kilmer, T. L. Quindry. 1978. Design guide for reducing transportation noise in and around buildings. Building Science Series 84, U. S. Department of Commerce, National Bureau of Standards, April.

27. Reynolds, D. D. 1981. Engineering principles of acoustics – noise and vibration control. Allyn and Bacon.

REFERENCES

28. Reynolds, D. D., J. M. Bledsoe. 1989. "Sound attenuation of acoustically lined circular ducts and radiused elbows." ASHRAE Transactions, Vol. 95, Part 1.

29. Reynolds, D. D., J. M. Bledsoe. 1989, "Sound attenuation of unlined and acoustically lined rectangular ducts." ASHRAE Transactions, Vol. 95, Part 1.

30. Reynolds, D. D., J. M. Bledsoe. 1989. "Sound transmission through mechanical equipment room walls, floor, or ceiling." ASHRAE Transactions, Vol. 95, Part 1.

31. Rudder, F. F., and S. L. Yaniv. 1985 "Guidelines for the prevention of traffic noise problems." NBSIR 84 – 2900 (DOT), Center for Building Technology, National Engineering Laboratory, National Bureau of Standards, April.

32. Sandbakken, M., L. Pande, M. J. Crocker. 1981. "Investigation of end reflection coefficient accuracy problems with AMCA Standard 300–67." HL 81–16, Ray W. Laboratories, April.

33. Schultz, T. J. 1985. "Relationship between sound power level and sound pressure level in dwellings and offices." ASHRAE Transactions, Vol. 91, Part 1.

34. Sessler, S. M. 1973. "Acoustical and mechanical considerations for the evaluation of chiller noise." ASHRAE Journal, October.

35. "Single–pass sound transmission loss data." 1981. Armstrong Cork Company.

36. "Sound and vibration control." Chapter 52, ASHRAE 1987 HVAC Systems and Applications Handbook.

37. "Sound and vibration control." Chapter 35, ASHRAE 1976 Systems Handbook.

38. "Sound and vibration control." Chapter 32, ASHRAE 1984 Systems Volume.

39. "Sound attenuation of fiberglass ducts and duct liners." 1970. Product Testing Laboratories Report No. 32433–Final, Owens–Corning Fiberglass Corporation, September 4.

40. "Sound control." Chapter 14, ASHRAE 1967 Guide and Data Book.

41. "Sound insertion loss properties of linacoustic and competitive duct liners." Report No. 436–T–1778, Johns–Manville Research and Development Center, December 19.

42. Thompson, J. K. 1981. "The room acoustics equation: Its limitation and potential." ASHRAE Transactions, Vol. 87, Part 2.

43. Thompson, J. K. 1976. "A modified room acoustics approach to determine sound pressure levels in irregularly proportioned factory spaces," Inter–Noise 76.

44. Ver, I. L. 1982. "A study to determine the noise generation and noise attenuation of lined and unlined duct fittings." Report No. 5092, Bolt, Beranek and Newman, Inc., July.

45. Ver, I. L. 1978. "A review of the attenuation of sound in straight lined and unlined ductwork of rectangular cross Section." ASHRAE Transactions, Vol. 84, Part 1.

46. Ver, I. L. 1984a. "Noise generation and noise attenuation of duct fittings—A review: Part II." ASHRAE Transactions, Vol. 90, Part 2A.

47. Ver, I. L. 1984b, "Prediction of sound transmission through duct walls: Breakout and pickup." ASHRAE Transactions, Vol. 90, Part. 2A.

48. Wells, R. J. 1958 "Acoustical plenum chambers." Noise Control, July.

49. AMCA. 1985. AMCA Standard 300–85, "Reverberent room method for sound testing of fans." Arlington Height, IL: Air Movement and Control Association, Inc.

50. ARI. 1987. ARI Standard 575–87, "Standard for method of measuring machinery sound within equipment rooms." Arlington, VA: Air Conditioning & Refrigeration Institute.

51. ARI 1987 .ARI Standard 880–87, "Industry standard for air terminals." Arlington, VA: Air Conditioning & Refrigeration Institute.

52. ASHRAE. 1986. ASHRAE Standard 68–1986, "Laboratory method of testing in–duct sound power measurement procedure for fans." Atlanta, GA: American Society of heating, Refrigerating, and Air–Conditioning Engineers, Inc.

53. ASTM. 1984. ASTM Standard E477–84, "Standard method of testing duct liner materials and prefabricated silencers for acoustical and airflow performance." Philadelphia, PA: American Society for Testing and Materials.

54. Personal correspondence from J. Barrie Graham, June 13, 1989.

APPENDIX 1

COMPUTER PROGRAMS FOR CHAPTER 2

ADDITION OF SOUND LEVELS

PROGRAM OUTLINE

Select addition of sound power or sound pressure levels
Input sound levels and insertion loss values
Add sound levels and insertion loss values
Print out results

COMPUTER PROGRAM – ADDSL.BAS

```
10  ' >ADDITION OF SOUND LEVELS - ADDSL.BAS<
20  '
30  DIM U%(8), T%(20, 8), S1(21)
40  '
50  '     List of Variables
60  '
70  '     FZ = 1/1 octave band center freqency, Hz
80  '     K = counter
90  '     S = selection for sound power or sound pressure levels
100 '     S1(K) = flag
110 '     T%() = array containing sound levels, dB
120 '     T$() = labels corresponding to output
130 '     U%() = 1/1 octave band sum sound power or sound pressure level, dB
140 '
150 CLS : PRINT "ADDITION OF SOUND POWER AND SOUND PRESSURE LEVELS": PRINT
160 PRINT "1/1 OCTAVE BAND DATA FOR FIRST DATA SET"
170 PRINT "1) SOUND POWER OR INTENSITY LEVELS"
180 PRINT "2) SOUND PRESSURE LEVELS"
190 INPUT "SELECTION: ", S
200 IF S > 2 OR S < 1 THEN GOTO 170 ELSE GOTO 210
210 K = 1: S1(1) = 2: ' Set flag for K = 1
220 '
230 '     Enter First Set of Sound Power/Intensity/Sound Pressure Levels
240 '
250 PRINT
260 PRINT "1/1 OCTAVE BAND DATA"
270 PRINT
280 FOR I = 1 TO 8
290     FZ = 62.5 * 2 ^ (I - 1): ' 1/1 Octave Band Center Frequency
300     PRINT "ENTER DATA FOR THE "; FZ; "HZ OCTAVE BAND"; : INPUT T%(1, I): U%(I) = T%(1, I)
310 NEXT I
320 PRINT "CONTINUE? (Y OR N)"
330 Q$ = INKEY$: IF Q$ = "Y" OR Q$ = "N" OR Q$ = "y" OR Q$ = "n" THEN GOTO 340 ELSE GOTO 330
340 IF Q$ = "N" OR Q$ = "n" THEN GOTO 690
350 '
360 GOSUB 1480: ' Enter Line Lable
370 '
380 K = 2
390 CLS
400 IF S = 2 THEN GOTO 560: ' S eqauls 2 for sound pressure levels
410 '
420 '     Sound Power Level Menu
430 '
440 PRINT "1. SUBTRACT INSERTION LOSS/NOISE REDUCTION/TRANSMISSION LOSS VALUES"
450 PRINT "2. ADD SOUND POWER OR INTENSITY LEVELS"
460 PRINT "3. TOTAL"
470 INPUT "SELECTION: ", S1(K)
480 '
490 IF S1(K) >= 1 AND S1(K) <= 3 THEN GOTO 500 ELSE GOTO 470
500 IF S1(K) = 3 THEN GOSUB 870
510 '
520 GOTO 660: ' Skips over menu for sound pressure levels
530 '
540 '     Sound Pressure Level Menu
550 '
560 PRINT "1. SUBTRACT INSERTION LOSS/NOISE REDUCTION VALUES"
570 PRINT "2. ADD SOUND PRESSURE LEVELS"
580 PRINT "3. TOTAL"
590 INPUT "SELECTION: ", S1(K)
```

```
600 PRINT
610 '
620 IF S1(K) >= 1 AND S1(K) <= 3 THEN GOTO 630 ELSE GOTO 590
630 IF S1(K) = 3 THEN GOSUB 870
640 '
650 '    Enter Subsequent Sets of Values
660 '
670 PRINT
680 PRINT "1/1 OCTAVE BAND DATA"
690 PRINT
700 '
710 FOR I = 1 TO 8
720    FZ = 62.5 * 2 ^ (I - 1): ' 1/1 Octave Band Center Frequency
730    PRINT "ENTER DATA FOR THE "; FZ; "HZ OCTAVE BAND"; : INPUT T%(K, I)
740 NEXT I
750 PRINT "CONTINUE? (Y OR N)"
760 Q$ = INKEY$: IF Q$ = "Y" OR Q$ = "N" OR Q$ = "y" OR Q$ = "n" THEN GOTO 770 ELSE GOTO 760
770 IF Q$ = "N" OR Q$ = "n" THEN GOTO 690
780 IF S1(K) = 1 THEN FOR I = 1 TO 8: T%(K, I) = -T%(K, I): NEXT I
790 '
800 GOSUB 1480: ' Enter Line Lable
810 '
820 K = K + 1
830 '
840 GOTO 390
850 '
860 '
870 CLS : LCNT = 1: FOR I = 1 TO 8: U%(I) = 0: NEXT I
880 PRINT SPC(48); "1/1 OCTAVE BAND CENTER FREQ"
890 PRINT SPC(13); "DESCRIPTION"; SPC(22); " 63  125  250  500   1K   2K   4K   8K"
900 B$ = STRING$(44, CHR$(45)) + "  ___ ___ ___ ___ ___ ___ ___ ___"
910 PRINT B$: LCNT = LCNT + 1
920 FOR I = 1 TO K
930    IF Q$ = "Y" THEN GOTO 980
940    IF S1(I) = 1 AND LCNT + 4 >= 23 THEN GOTO 960 ELSE GOTO 950
950    IF S1(I) = 2 AND LCNT + 4 >= 21 THEN GOTO 960 ELSE GOTO 980
960    PRINT TAB(54); ">PRESS RETURN TO CONTINUE<"
970    QQ$ = INKEY$: IF QQ$ <> CHR$(13) THEN GOTO 970 ELSE LCNT = 1
980    IF S1(I) = 1 THEN GOSUB 1040
990    IF S1(I) = 2 THEN GOSUB 1170
1000 NEXT I
1010 '
1020 END
1030 '
1040 '    >Addition of IL, NR, TL subroutine<
1050 '
1060 FOR J = 1 TO 8
1070    U%(J) = U%(J) + T%(I, J)
1080    IF U%(J) < 5 THEN U%(J) = 5
1090 NEXT J
1100 GOSUB 1380
1110 IF I < K - 1 THEN GOTO 1150
1120 PRINT B$
1130 GOSUB 1440
1140 PRINT B$
1150 RETURN
1160 '
1170 '    >Addition of Lw subroutine<
1180 '
1190 IF K > 1 AND S1(I - 1) = 2 THEN GOTO 1230
1200 IF I = 1 THEN GOTO 1060
1210 PRINT B$: LCNT = LCNT + 1
1220 GOSUB 1440: LCNT = LCNT + 1
1230 GOSUB 1380
1240 IF I <= K - 2 AND S1(I + 1) = 1 THEN PRINT B$: LCNT = LCNT + 1
1250 IF I = K - 1 THEN PRINT B$
1260 FOR J = 1 TO 8
1270    UA = 10 ^ (U%(J) / 10): UB = 10 ^ (T%(I, J) / 10): UC = UA + UB
1280    UC = 10 * LOG(UC) / LOG(10): U%(J) = UC
1290 NEXT J
1300 IF I <= K - 2 AND S1(I + 1) = 2 THEN I = I + 1: GOTO 1230
1310 IF S = 1 THEN PRINT SPC(8); "SUM SOUND POWER LEVELS"; SPC(15);
1320 IF S = 2 THEN PRINT SPC(8); "SUM SOUND PRESSURE LEVELS"; SPC(12);
1330 PRINT USING " ###"; U%(1); U%(2); U%(3); U%(4); U%(5); U%(6); U%(7); U%(8)
1340 IF I < K - 1 THEN LCNT = LCNT + 1
```

```
1350 IF I = K - 1 THEN PRINT B$
1360 RETURN
1370 '
1380 PRINT USING "## "; I;
1390 PRINT USING "\                                          \"; T$(I);
1400 PRINT USING " ###"; T%(I, 1); T%(I, 2); T%(I, 3); T%(I, 4); T%(I, 5); T%(I, 6); T%(I, 7); T%(I, 8)
1410 LCNT = LCNT + 1
1420 RETURN
1430 '
1440 PRINT SPC(5); "SUM WITH NOISE REDUCTION VALUES"; SPC(9);
1450 PRINT USING " ###"; U%(1); U%(2); U%(3); U%(4); U%(5); U%(6); U%(7); U%(8)
1460 RETURN
1470 '
1480 '    >Line Lable Subroutine<
1490 '
1500 PRINT
1510 PRINT "ENTER LINE LABLE:"
1520 PRINT STRING$(44, CHR$(45))
1530 LINE INPUT T$(K)
1540 PRINT "CONTINUE? (Y OR N)"
1550 Q$ = INKEY$: IF Q$ = "N" OR Q$ = "Y" OR Q$ = "y" OR Q$ = "n" THEN GOTO 1560 ELSE GOTO 1550
1560 IF Q$ = "N" OR Q$ = "n" THEN GOTO 1500
1570 RETURN
```

SAMPLE OUTPUT FOR EXAMPLE 2.1

ADDITION OF SOUND POWER AND SOUND PRESSURE LEVELS

1/1 OCTAVE BAND DATA FOR FIRST DATA SET
1) SOUND POWER OR INTENSITY LEVELS
2) SOUND PRESSURE LEVELS
SELECTION: 1

1/1 OCTAVE BAND DATA

```
ENTER DATA FOR THE   62.5 HZ OCTAVE BAND? 84
ENTER DATA FOR THE   125 HZ OCTAVE BAND? 87
ENTER DATA FOR THE   250 HZ OCTAVE BAND? 90
ENTER DATA FOR THE   500 HZ OCTAVE BAND? 89
ENTER DATA FOR THE   1000 HZ OCTAVE BAND? 91
ENTER DATA FOR THE   2000 HZ OCTAVE BAND? 88
ENTER DATA FOR THE   4000 HZ OCTAVE BAND? 87
ENTER DATA FOR THE   8000 HZ OCTAVE BAND? 84
CONTINUE? (Y OR N)
```

ENTER LINE LABLE: _____

SOUND POWER LEVEL — LW1
CONTINUE (Y OR N)

1. SUBTRACT INSERTION LOSS/NOISE REDUCTION/TRANSMISSION LOSS VALUES
2. ADD SOUND POWER LEVELS
3. TOTAL
SELECTION: 2

1/1 OCTAVE BAND DATA

```
ENTER DATA FOR THE   62.5 HZ OCTAVE BAND? 74
ENTER DATA FOR THE   125 HZ OCTAVE BAND? 95
ENTER DATA FOR THE   250 HZ OCTAVE BAND? 93
ENTER DATA FOR THE   500 HZ OCTAVE BAND? 98
ENTER DATA FOR THE   1000 HZ OCTAVE BAND? 99
ENTER DATA FOR THE   2000 HZ OCTAVE BAND? 92
ENTER DATA FOR THE   4000 HZ OCTAVE BAND? 88
ENTER DATA FOR THE   8000 HZ OCTAVE BAND? 87
CONTINUE? (Y or N)
```

ENTER LINE LABLE: _____

SOUND POWER LEVEL — LW1
CONTINUE (Y OR N)

1. SUBTRACT INSERTION LOSS/NOISE REDUCTION/TRANSMISSION LOSS VALUES
2. ADD SOUND POWER LEVELS
3. TOTAL
SELECTION: 2

1/1 OCTAVE BAND DATA

```
ENTER DATA FOR THE   62.5 HZ OCTAVE BAND? 70
ENTER DATA FOR THE   125 HZ OCTAVE BAND? 78
ENTER DATA FOR THE   250 HZ OCTAVE BAND? 86
ENTER DATA FOR THE   500 HZ OCTAVE BAND? 86
ENTER DATA FOR THE   1000 HZ OCTAVE BAND? 73
ENTER DATA FOR THE   2000 HZ OCTAVE BAND? 67
ENTER DATA FOR THE   4000 HZ OCTAVE BAND? 59
ENTER DATA FOR THE   8000 HZ OCTAVE BAND? 57
CONTINUE? (Y or N)
```

ENTER LINE LABLE: _____

SOUND POWER LEVEL — LW1
CONTINUE (Y OR N)

1. SUBTRACT INSERTION LOSS/NOISE REDUCTION/TRANSMISSION LOSS VALUES
2. ADD SOUND POWER LEVELS
3. TOTAL
SELECTION: 1

1/1 OCTAVE BAND DATA

ENTER DATA FOR THE   62.5 HZ OCTAVE BAND? 6
ENTER DATA FOR THE   125 HZ OCTAVE BAND? 11
ENTER DATA FOR THE   250 HZ OCTAVE BAND? 17
ENTER DATA FOR THE   500 HZ OCTAVE BAND? 22
ENTER DATA FOR THE   1000 HZ OCTAVE BAND? 28
ENTER DATA FOR THE   2000 HZ OCTAVE BAND? 32
ENTER DATA FOR THE   4000 HZ OCTAVE BAND? 24
ENTER DATA FOR THE   8000 HZ OCTAVE BAND? 20
CONTINUE? (Y or N)

ENTER LINE LABLE:
_____
SOUND POWER LEVEL — LW1
CONTINUE (Y OR N)

1. SUBTRACT INSERTION LOSS/NOISE REDUCTION/TRANSMISSION LOSS VALUES
2. ADD SOUND POWER LEVELS
3. TOTAL
SELECTION: 3

		1/1 OCTAVE BAND CENTER FREQ						
DESCRIPTION	63	125	250	500	1K	2K	4K	8K
1 SOUND POWER LEVEL — LW1	84	87	90	89	91	88	87	84
2 SOUND POWER LEVEL — LW2	74	95	93	98	99	92	88	87
3 SOUND POWER LEVEL — LW2	70	78	86	86	73	67	59	57
SUM SOUND POWER LEVELS	84	96	96	99	100	93	91	89
4 INSERTION LOSS — IL	−6	−11	−17	−22	−28	−32	−24	−20
SUM WITH NOISE REDUCTION VALUES	78	85	79	77	72	61	67	69

SAMPLE OUTPUT FOR EXAMPLE 2.2

ADDITION OF SOUND POWER AND SOUND PRESSURE LEVELS

1/1 OCTAVE BAND DATA FOR FIRST DATA SET
1) SOUND POWER OR OR INTENSITY LEVELS
2) SOUND PRESSURE LEVELS
SELECTION: 2

1/1 OCTAVE BAND DATA

ENTER DATA FOR THE   62.5 HZ OCTAVE BAND? 74
ENTER DATA FOR THE   125 HZ OCTAVE BAND? 78
ENTER DATA FOR THE   250 HZ OCTAVE BAND? 77
ENTER DATA FOR THE   500 HZ OCTAVE BAND? 77
ENTER DATA FOR THE   1000 HZ OCTAVE BAND? 75
ENTER DATA FOR THE   2000 HZ OCTAVE BAND? 77
ENTER DATA FOR THE   4000 HZ OCTAVE BAND? 75
ENTER DATA FOR THE   8000 HZ OCTAVE BAND? 66
CONTINUE? (Y or N)

ENTER LINE LABLE:
_____
SOUND PRESSURE LEVEL — LP1
CONTINUE (Y OR N)

1. SUBTRACT INSERTION LOSS/NOISE REDUCTION VALUES
2. ADD SOUND PRESSURE LEVELS
3. TOTAL
SELECTION: 2

1/1 OCTAVE BAND DATA

```
ENTER DATA FOR THE    62.5 HZ OCTAVE BAND? 73
ENTER DATA FOR THE    125  HZ OCTAVE BAND? 83
ENTER DATA FOR THE    250  HZ OCTAVE BAND? 86
ENTER DATA FOR THE    500  HZ OCTAVE BAND? 80
ENTER DATA FOR THE    1000 HZ OCTAVE BAND? 76
ENTER DATA FOR THE    2000 HZ OCTAVE BAND? 78
ENTER DATA FOR THE    4000 HZ OCTAVE BAND? 79
ENTER DATA FOR THE    8000 HZ OCTAVE BAND? 71
CONTINUE? (Y or N)
```

ENTER LINE LABLE:
_____

SOUND PRESSURE LEVEL — LW2
CONTINUE (Y OR N)

1. SUBTRACT INSERTION LOSS/NOISE REDUCTION VALUES
2. ADD SOUND PRESSURE LEVELS
3. TOTAL
SELECTION: 2

1/1 OCTAVE BAND DATA

```
ENTER DATA FOR THE    62.5 HZ OCTAVE BAND? 84
ENTER DATA FOR THE    125  HZ OCTAVE BAND? 89
ENTER DATA FOR THE    250  HZ OCTAVE BAND? 88
ENTER DATA FOR THE    500  HZ OCTAVE BAND? 86
ENTER DATA FOR THE    1000 HZ OCTAVE BAND? 89
ENTER DATA FOR THE    2000 HZ OCTAVE BAND? 88
ENTER DATA FOR THE    4000 HZ OCTAVE BAND? 85
ENTER DATA FOR THE    8000 HZ OCTAVE BAND? 78
CONTINUE? (Y or N)
```

ENTER LINE LABLE:
_____

SOUND PRESSURE LEVEL — LP3
CONTINUE (Y OR N)

1. SUBTRACT INSERTION LOSS LEVELS
2. ADD SOUND PRESSURE LEVELS
3. TOTAL
SELECTION: 3

DESCRIPTION	1/1 OCTAVE BAND CENTER FREQ							
	63	125	250	500	1K	2K	4K	8K
1 SOUND PRESSURE LEVEL — LP1	74	78	77	77	75	77	75	66
2 SOUND PRESSURE LEVEL — LP2	73	83	86	80	76	78	79	71
3 SOUND PRESSURE LEVEL — LP3	84	89	88	86	89	88	85	78
SUM SOUND PRESSURE LEVELS	85	90	91	87	89	89	86	79

NOISE CRITERION CURVES

PROGRAM OUTLINE

Input System Parameters:
    1/1 octave band sound pressure levels, dB
    NC curve comparison or RC curve calculation
Calculate:
    NC curve comparison curve and subjective criteria
Print out results

COMPUTER PROGRAM – NCLEVEL.BAS

```
10  '     >SUBROUTINE TO COMPARE PL'S WITH NC LEVEL - NCLEVEL.BAS<
20  '
30  DIM NC%(11, 8), EF(11), U%(8), TT%(8)
40  '
50  '     List of Variables
60  '
70  '     EF() = NC-curve designations
80  '     NC%() = values for NC curves, dB
90  '     R() = value used for interpolation
100 '     TT%() = values for desired NC curve
110 '     U%() = Sound pressure level, dB
120 '
130 FOR I = 1 TO 11
140   FOR J = 1 TO 8
150     READ NC%(I, J): 'Reads in values for NC curves
160   NEXT J
170 NEXT I
180 '
190 CLS : PRINT "NOISE CRITERION (NC) CALCULATIONS": PRINT
200 '
210 INPUT "SOUND PRESSURE LEVEL FOR 63 Hz"; U%(1)
220 INPUT "SOUND PRESSURE LEVEL FOR 125 Hz"; U%(2)
230 INPUT "SOUND PRESSURE LEVEL FOR 250 Hz"; U%(3)
240 INPUT "SOUND PRESSURE LEVEL FOR 500 Hz"; U%(4)
250 INPUT "SOUND PRESSURE LEVEL FOR 1K Hz"; U%(5)
260 INPUT "SOUND PRESSURE LEVEL FOR 2K Hz"; U%(6)
270 INPUT "SOUND PRESSURE LEVEL FOR 4K Hz"; U%(7)
280 INPUT "SOUND PRESSURE LEVEL FOR 8K Hz"; U%(8)
290 '
300 PRINT "CONTINUE? (Y/N)"
310 Q$ = INKEY$: IF Q$ = "Y" OR Q$ = "N" OR Q$ = "y" OR Q$ = "n" THEN GOTO 320 ELSE GOTO 310
320 IF Q$ = "N" OR Q$ = "n" THEN GOTO 190
330 '
340 '     >COMPARION OF NC LEVELS<
350 '
360 FOR I = 1 TO 11
370   EF(I) = 70 - 5 * I
380 NEXT I
390 '
400 MAX = 0: FOR I = 1 TO 8: R(I) = 0: NEXT I
410 FOR I = 1 TO 8
420   IF U%(I) > NC%(1, I) THEN NC$ = "NC > 65": PRINT : GOTO 830
430 NEXT I
440 '
450 FOR J = 1 TO 8
460   FOR I = 1 TO 11
470     IF U%(J) = NC%(I, J) THEN R(J) = 70 - 5 * I: GOTO 520
480     IF U%(J) < NC%(I, J) THEN GOTO 510
490     IF U%(J) > NC%(I, J) THEN X = (NC%(I - 1, J) - U%(J)) / (NC%(I - 1, J) - NC%(I, J))
500     R(J) = 70 - 5 * (I - 1) - 5 * X: GOTO 520
510   NEXT I
520 NEXT J
530 '
540 FOR I = 1 TO 8
550   IF MAX < R(I) THEN MAX = R(I)
560 NEXT I
```

```
570 '
580 IF MAX = 0 THEN NC$ = "NC < 15": PRINT : GOTO 830
590 MAX = CINT(MAX)
600 NC$ = "NC" + STR$(MAX)
610 '
620 RR = INT(1 + (65 - MAX) / 5): IF MAX = EF(RR) THEN GOTO 630 ELSE GOTO 640
630 FOR I = 1 TO 8: TT%(I) = NC%(RR, I): NEXT I: GOTO 710
640 FOR I = 1 TO 8
650     TT%(I) = (NC%(RR, I) - NC%(RR + 1, I)) * (EF(RR) - MAX) / 5
660     TT%(I) = NC%(RR, I) - TT%(I)
670 NEXT I
680 '
690 ' Output Statements
700 '
710 PRINT " "
720 PRINT "                    1/1 OCT BAND CENTER FREQUENCIES"
730 PRINT SPC(14); "63    125    250    500    1K    2K    4K    8K"
740 PRINT STRING$(59, CHR$(196))
750 PRINT "NC"; MAX; "CURVE   ";
760 PRINT USING "###    "; TT%(1); TT%(2); TT%(3); TT%(4); TT%(5); TT%(6); TT%(7); TT%(8)
770 PRINT "LEVELS:       ";
780 PRINT USING "###    "; U%(1); U%(2); U%(3); U%(4); U%(5); U%(6); U%(7); U%(8)
790 PRINT STRING$(59, CHR$(196))
800 PRINT "DIFF. :       ";
810 PRINT USING "###    "; U%(1) - TT%(1); U%(2) - TT%(2); U%(3) - TT%(3); U%(4) - TT%(4); U%(5) - TT%(5); U%(6) - TT%(6); U%(7) - TT%(7); U%(8) - TT%(8)
820 PRINT : PRINT :
830 PRINT "THE NC RATING IS "; NC$
840 END
850 '
860 ' Data for NC-curves: 63 - 8K Hz Octave bands
870 '
880 DATA 80,75,71,68,66,64,63,62: ' Data for NC-65
890 DATA 77,71,67,63,61,59,58,57: ' Data for NC-60
900 DATA 74,67,62,58,56,54,53,52: ' Data for NC-55
910 DATA 71,64,58,54,51,49,48,47: ' Data for NC-50
920 DATA 67,60,54,49,46,44,43,42: ' Data for NC-45
930 DATA 64,56,50,45,41,39,38,37: ' Data for NC-40
940 DATA 60,52,45,40,36,34,33,32: ' Data for NC-35
950 DATA 57,48,41,35,31,29,28,27: ' Data for NC-30
960 DATA 54,44,37,31,27,24,22,21: ' Data for NC-25
970 DATA 51,40,33,26,22,19,17,16: ' Data for NC-20
980 DATA 47,36,29,22,17,14,12,11: ' Data for NC-15
```

SAMPLE OUTPUT FOR EXAMPLE 2.3

NOISE CRITERION CURVES

```
SOUND PRESSURE LEVEL FOR 63 Hz? 50
SOUND PRESSURE LEVEL FOR 125 Hz? 55
SOUND PRESSURE LEVEL FOR 250 Hz? 58
SOUND PRESSURE LEVEL FOR 500 Hz? 58
SOUND PRESSURE LEVEL FOR 1K Hz? 55
SOUND PRESSURE LEVEL FOR 2K Hz? 50
SOUND PRESSURE LEVEL FOR 4K Hz? 45
SOUND PRESSURE LEVEL FOR 8K Hz? 39
CONTINUE? (Y OR N)
```

	1/1 OCT BAND CENTER FREQUENCIES							
	63	125	250	500	1K	2K	4K	8K
NC 55 CURVE	74	67	62	58	56	54	53	52
LEVELS:	50	55	58	58	55	50	45	39
DIFF.:	−24	−12	−4	0	−1	−4	−8	−13

THE NC RATING IS   55

ROOM CRITERION CURVES

PROGRAM OUTLINE

Input System Parameters:
    1/1 octave band sound pressure levels, dB
    RC curve calculation
Calculate:
    RC curve comparison curve and subjective criteria
Print out results

COMPUTER PROGRAM – RCLEVEL.BAS

```
10  '    >SUBROUTINE TO COMPARE PL'S WITH RC LEVEL - RCLEVEL.BAS<
20  '
30  DIM U%(8), TT%(8), RC$(8)
40  '
50  '    List of Variables
60  '
70  '    A$ = Stores or "RC" for use in output statements
80  '    CRIT = Value of RC level, dB
90  '    Q$ = For use in question for continuing the program or not
100 '    TT%() = Values or RC curve, dB
110 '    U%() = Sound pressure level, dB
120 '
130 CLS : PRINT "ROOM CRITERION (RC) CALCULATIONS": PRINT
140 '
150 INPUT "SOUND PRESSURE LEVEL FOR 31.5 Hz"; U%(0)
160 INPUT "SOUND PRESSURE LEVEL FOR 63 Hz"; U%(1)
170 INPUT "SOUND PRESSURE LEVEL FOR 125 Hz"; U%(2)
180 INPUT "SOUND PRESSURE LEVEL FOR 250 Hz"; U%(3)
190 INPUT "SOUND PRESSURE LEVEL FOR 500 Hz"; U%(4)
200 INPUT "SOUND PRESSURE LEVEL FOR 1K Hz"; U%(5)
210 INPUT "SOUND PRESSURE LEVEL FOR 2K Hz"; U%(6)
220 INPUT "SOUND PRESSURE LEVEL FOR 4K Hz"; U%(7)
230 '
240 PRINT "CONTINUE? (Y/N)"
250 Q$ = INKEY$: IF Q$ = "Y" OR Q$ = "N" OR Q$ = "y" OR Q$ = "n" THEN GOTO 260 ELSE GOTO 250
260 IF Q$ = "N" OR Q$ = "n" THEN GOTO 130
270 '
280 '    >COMPARISON WITH ROOM CRITERION CURVES<
290 '
300 TT%(5) = (U%(4) + U%(5) + U%(6)) / 3: ' Determines RC level
310 CRIT = TT%(5): A$ = "RC": ' For use with output
320 FOR I = 0 TO 7
330    TT%(I) = TT%(5) + (5 - I) * 5
340    IF I < 5 AND U%(I) > TT%(I) + 5 THEN RUMBLE = 1: ' Checks for Rumble
350    IF I > 4 AND U%(I) > TT%(I) + 3 THEN HISS = 1: ' Checks for Hiss
360    IF I = 0 THEN GOTO 380
370    IF U%(I) > U%(I - 1) + 5 THEN TONAL = 1: ' Checks for Tonal
380    FZ = 62.5 * 2 ^ (I - 1)
390    LOGFZ = LOG(FZ) / LOG(10)
400    X = 13.288 * LOG(FZ / 31.5) / LOG(10)
410    IF U%(I) > X + 65 THEN FLAG = 1: ' Checks to see if Lp is in region B
420    IF U%(I) > X + 75 THEN FLAG2 = 1: ' Checks to see if Lp is in region A
430 NEXT I
440 '
450 IF RUMBLE = 1 THEN RC$ = "R"
460 IF HISS = 1 THEN RC$ = "H"
470 IF RUMBLE = 1 AND HISS = 1 THEN RC$ = "RH"
480 IF TONAL = 1 THEN RC$ = "T"
490 IF RUMBLE = 1 AND TONAL = 1 THEN RC$ = "R T"
500 IF HISS = 1 AND TONAL = 1 THEN RC$ = "H T"
510 IF RUMBLE = 1 AND HISS = 1 AND TONAL = 1 THEN RC$ = "RH T"
520 IF RC$ = "" THEN RC$ = "N": ' Neutral
530 '
540 IF FLAG = 1 THEN REG$ = "POSSIBILITY OF NOISE-INDUCED VIBRATION"
550 IF FLAG2 = 1 THEN REG$ = "HIGH PROBABILITY OF NOISE-INDUCED VIBRATION"
560 '
```

```
570 '   Output Statements
580 '
590 PRINT
600 PRINT SPC(14); "1/1 OCT BAND CENTER FREQUENCIES"
610 PRINT SPC(10); "31.5   63   125   250   500   1K   2K   4K"
620 PRINT STRING$(55, CHR$(45))
630 PRINT "LEVELS:";
640 PRINT USING "###   "; U%(0); U%(1); U%(2); U%(3); U%(4); U%(5); U%(6); U%(7)
650 PRINT A$; "-"; CRIT; ":";
660 PRINT USING "###   "; TT%(0); TT%(1); TT%(2); TT%(3); TT%(4); TT%(5); TT%(6); TT%(7)
670 PRINT STRING$(55, CHR$(45))
680 PRINT "DIFF. :";
690 PRINT USING "###   "; U%(0) - TT%(0); U%(1) - TT%(1); U%(2) - TT%(2); U%(3) - TT%(3); U%(4) - TT%(4); U%(5) - TT%(5); U%(6) - TT%(6); U%(7) - TT%(7)
700 IF FLAG = 1 THEN PRINT : PRINT REG$
710 PRINT
720 PRINT "THE RC LEVEL IS RC"; CRIT; "("; RC$; ")"
730 PRINT : PRINT :
740 PRINT "NEGATIVE NUMBERS ARE BLEOW THE RC -"; CRIT; "CURVE"
750 END
```

SAMPLE OUTPUT FOR EXAMPLE 2.4

ROOM CRITERION (RC) CURVES

```
SOUND PRESSURE LEVEL FOR 31.5 Hz? 70
SOUND PRESSURE LEVEL FOR 63 Hz? 62
SOUND PRESSURE LEVEL FOR 125 Hz? 54
SOUND PRESSURE LEVEL FOR 250 Hz? 46
SOUND PRESSURE LEVEL FOR 500 Hz? 40
SOUND PRESSURE LEVEL FOR 1K Hz? 33
SOUND PRESSURE LEVEL FOR 2K Hz? 27
SOUND PRESSURE LEVEL FOR 4K Hz? 20
CONTINUE? (Y OR N)
```

	1/1 OCT BAND CENTER FREQUENCIES							
	31.5	63	125	250	500	1K	2K	4K
LEVELS:	70	62	54	46	40	33	27	20
RC- 33 :	58	53	48	43	38	33	28	23
DIFF. :	12	9	6	3	2	0	-1	-3

POSSIBILITY OF NOISE-INDUCED VIBRATION

THE RC LEVEL IS 33 (R)

NEGATIVE NUMBERS ARE BELOW THE RC - 33 CURVE.

COMPUTER PROGRAMS FOR CHAPTER 3

APPENDIX 2

COMPUTER PROGRAMS FOR CHAPTER 3   A2–1

FANS

PROGRAM OUTLINE

Input data from Table 3.1; Specific sound power levels
Input data from Table 3.2; Blade frequency increments
Input data from Table 3.3; Correction factor for off–peak operation
Input system parameters:
    Type of fan
    Fan volume flow rate, cfm
    Total static pressure, in H_2O
    Number of fan blades
    Fan speed, rpm
    Fan horsepower
    Fan peak efficiency
Calculate:
    Blade frequency increment (Table 3.2)
    Correction for off–peak operation (Table 3.3)
    Sound power level, dB

COMPUTER PROGRAM – FANPWL.BAS

```
10  ' >FAN SOUND POWER LEVEL COMPUTER PROGRAM<
20  '
30  DIM T%(8), D(12, 8), BP(12), BF(12)
40  '
50  '     LIST OF VARIABLES
60  '
70  '     BF() = Corresponding 1/1 Octave Band for BP() data (e.g. 1 = 63 Hz)
80  '     BP() = Blade Frequency Increment, dB (Table 2)
90  '     D = Correction factor for off peak operation (Table 3.3)
100 '     D() = Specific Sound Power Levels, db (Table 1)
110 '     E = Percentage of peak frequency, %
120 '     E1 = Fan efficiency at point of operation, %
130 '     E2 = Fan peak efficiency, %
140 '     FC = Blade passage frequency, Hz
150 '     FV = Fan volume flow rate, cfm
160 '     HP = Fan Horsepower, Hp
170 '     K = Correction for Blade Frequency Increment, dB
180 '     RPM = Fan Speed, rpm
190 '     SP = Total static pressure, in H20
200 '     S1 = Selection for type of fan
210 '     S2 = Selection for method of determining blade frequency increment
220 '     T%() = Sound power level output, dB
230 '
240 ' Read in data
250 '
260 FOR I = 1 TO 12
270  FOR J = 1 TO 8
280    READ D(I, J)
290  NEXT J
300 NEXT I
310 '
320 ' Data from Table 3.1
330 '
340 DATA 40,40,39,34,30,23,19,17
350 DATA 45,45,43,39,34,28,24,24
360 DATA 53,53,43,36,36,31,26,21
370 DATA 56,47,43,39,37,32,30,26
380 DATA 58,54,45,42,38,33,29,26
390 DATA 61,58,53,48,46,44,41,38
400 DATA 49,43,43,48,47,45,38,34
410 DATA 49,43,46,43,41,36,30,28
420 DATA 53,52,51,51,49,47,43,40
```

```
430 DATA 51,46,47,49,47,46,39,37
440 DATA 48,47,49,53,52,51,43,40
445 DATA 48,51,58,56,55,52,46,42
450 ' Reads in data from Table 3.2
460 FOR I = 1 TO 12
470 READ BP(I), BF(I)
480 NEXT I
490 DATA 3,3
500 DATA 3,3
510 DATA 2,4
520 DATA 8,2
530 DATA 8,2
540 DATA 8,2
545 DATA 6,2
550 DATA 6,2
560 DATA 6,2
570 DATA 7,1
580 DATA 7,1
590 DATA 5,1
600 '
610 ' >Fan PWL Level data subroutine<
620 ' ASHRAE method for generalozed total fan sound power
630 REM >ASHRAE fan PWL subroutine<
640 CLS
650 PRINT "ASHRAE FAN SOUND POWER LEVEL MENU"
660 PRINT
670 PRINT "CENTRIFUGAL FANS"
680 PRINT " 1. AIRFOIL/BACKWARD CURVED - >36 IN DIA"
690 PRINT " 2. AIRFOIL/BACKWARD CURVED - <36 IN DIA"
700 PRINT " 3. FORWARD CURVED - ALL FANS
710 PRINT " 4. RADIAL - MATERIAL WHEEL - TOTAL PRESSURE: 4-12 IN H20"
720 PRINT " 5. RADIAL - MEDIUM PRESSURE - TOTAL PRESSURE: 8-15 IN H20"
730 PRINT " 6. RADIAL - HIGH PRESSURE - TOTAL PRESSURE: 15-40 IN H20": PRINT
740 PRINT "VANEAXIAL FANS"
750 PRINT " 7. HUB RATIO - 0.3-0.4"
760 PRINT " 8. HUB RATIO - 0.4-0.6"
765 PRINT " 9. HUB RATIO - 0.6-0.8": PRINT
770 PRINT "TUBEAXIAL FANS"
780 PRINT "10. DIA >40 IN"
790 PRINT "11. DIA <40 IN": PRINT
800 PRINT "PROPELLER FANS"
810 PRINT "12. GENERALIZED VENTILATION AND COOLING TOWER": PRINT
820 INPUT "SELECTION: ", S1
830 IF S1 >= 1 AND S1 <= 12 THEN GOTO 840 ELSE GOTO 820
840 PRINT
850 INPUT "ENTER FAN FLOW VOLUME (CFM): ", FV
860 INPUT "ENTER FAN TOTAL STATIC PRESSURE (IN H20): ", SP
870 PRINT
880 PRINT "1.   NUMBER OF BLADES AND RPM AVAILABLE"
890 PRINT "2.   NUMBER OF BLADES AND RPM NOT AVAILABLE"
900 INPUT "SELECTION: ", S2
910 IF S2 >= 1 AND S2 <= 2 THEN GOTO 930 ELSE GOTO 900
920 IF S2 = 2 THEN GOTO 970
930 PRINT
940 INPUT "ENTER NUMBER OF BLADES: ", NB
950 INPUT "ENTER FAN SPEED (RPM): ", RPM
960 BFI = NB * RPM / 60
970 INPUT "ENTER FAN HORSEPOWER: ", HP
980 E1 = FV * SP / (6356 * HP): ' Fan Efficiency at point of operation
990 E1 = CINT(100 * E1)
1000 IF E1 > 100 THEN PRINT "FAN EFFICIENCY IS OVER 100%, MUST REENTER DATA": GOTO 840
1010 PRINT
1020 PRINT "FAN EFFICIENCY AT POINT OF OPERATION = "; E1; "%"
1030 PRINT "PEAK FAN EFFICENCY MUST BE GREATER THAN OR EQUAL TO"; E1; "%"
1040 PRINT "IF EFFICIENCY IS NOT KNOWN, ENTER"; E1; "%"
1050 INPUT "ENTER FAN PEAK EFFICIENCY (WHOLE NUMBER): ", E2
1060 PRINT
1070 PRINT "CONTINUE? (Y OR N)"
1080 Q$ = INKEY$: IF Q$ = "Y" OR Q$ = "N" OR Q$ = "y" OR Q$ = "n" THEN GOTO 1090 ELSE GOTO 1080
1090 IF Q$ = "N" OR Q$ = "n" THEN GOTO 640
1100 FC = (LOG(FV) / LOG(10)) * 10 + (LOG(SP) / LOG(10)) * 20: ' Equation 3.2 without the first term
1110 E = E1 / E2: E = CINT(100 * E): ' Efficiency for use with Table 3.3
1120 IF E > 100 THEN INPUT "EFFICIENCY IS OVER 100% MUST REENTER DATA", X$: GOTO 640
1130 '
1140 ' Correction for Off-Peak Operation (See Table 3.3)
```

```
1150 '
1160 IF E >= 90 AND E <= 100 THEN D = 0
1170 IF E >= 85 AND E < 90 THEN D = 3
1180 IF E >= 75 AND E < 85 THEN D = 6
1190 IF E >= 65 AND E < 75 THEN D = 9
1200 IF E >= 55 AND E < 65 THEN D = 12
1210 IF E >= 50 AND E < 55 THEN D = 15
1220 IF E < 50 THEN D = 16
1230 IF D > 15 THEN CLS : LOCATE 11, 23: PRINT "EFFICIENCY CORRECTION IS +": D: " DB."
1240 IF D > 15 THEN LOCATE 13, 19: PRINT "DO YOU WANT TO CHANGE FAN PARAMETERS? (Y/N) "
1250 IF D > 15 THEN Q3$ = INKEY$: IF Q3$ = "Y" OR Q3$ = "N" OR Q3$ = "y" OR Q3$ = "n" THEN GOTO 1260 ELSE GOTO 1240
1260 IF D > 15 AND (Q3$ = "Y" OR Q3$ = "y") THEN GOTO 630
1270 FOR I = 1 TO 8
1280 T%(I) = D(S1, I) + FC + D: ' Equation 3.2 inclusive
1290 NEXT I
1300 IF S2 = 2 THEN K = BF(S1)
1310 IF S2 = 2 AND RPM > 1750 THEN K = K + 1
1320 IF S2 = 1 THEN K = CINT(3.3219 * LOG(BFI / 31.25) / LOG(10))
1330 T%(K) = T%(K) + BP(S1): ' Adds Blade Frequency Increment
1340 '
1350 ' Output Statements
1360 '
1370 PRINT SPC(45); "1/1 OCT BAND CENTER FREQ"
1380 PRINT SPC(10); "DESCRIPTION"; SPC(21); " 63 125 250 500  1K   2K   4K   8K"
1390 PRINT STRING$(41, CHR$(45)) + " ___ ___ ___ ___ ___ ___ ___ ___"
1400 PRINT USING "\                                        \"; "FAN SOUND POWER LEVELS, dB";
1410 PRINT USING " ###"; T%(1); T%(2); T%(3); T%(4); T%(5); T%(6); T%(7); T%(8)
1420 END
```

SAMPLE OUTPUT FOR EXAMPLE 3.1

```
ASHRAE FAN SOUND POWER LEVEL MENU

CENTRIFUGAL FANS
1. AIRFOIL/BACKWARD CURVED — >36 IN DIA
2. AIRFOIL/BACKWARD CURVED — <36 IN DIA
3. FORWARD CURVED — ALL FANS
4. RADIAL — MATERIAL WHEEL — TOTAL PRESSURE: 4—10 IN H2O
5. RADIAL — MEDIUM PRESSURE — TOTAL PRESSURE: 6—15 IN H2O
6. RADIAL — HIGH PRESSURE — TOTAL PRESSURE: 15—60 IN H2O

VANEAXIAL FANS
7. HUB RATIO — 0.3—0.4
8. HUB RATIO — 0.4—0.6
9. HUB RATIO — 0.6—0.8

TUBEAXIAL FANS
10. DIA >40 IN
11. DIA <40 IN

PROPELLER FANS
12. GENERAL VENTILATION AND COOLING TOWER

SELECTION: 3

ENTER FAN FLOW VOLUME (CFM): 10000
ENTER FAN TOTAL STATIC PRESSURE (IN H2O): 1.5

1. NUMBER OF BLADES AND RPM AVAILABLE
2. NUMBER OF BLADES AND RPM NOT AVAILABLE
SELECTION: 1

ENTER NUMBER OF BLADES: 24
ENTER FAN SPEED (RPM): 1175
ENTER FAN HORSPOWER: 3

FAN EFFICIENCY AT POINT OF OPERATION =  79 %
PEAK FAN EFFICENCY MUST BE GREATER THAN OR EQUAL TO 79 %
IF EFFICIENCY IS NOT KNOWN, ENTER 79 %
ENTER FAN PEAK EFFICIENCY (WHOLE NUMBER):  85

CONTINUE? (Y OR N)
```

DESCRIPTION	\multicolumn{8}{c}{1/1 OCT BAND CENTER FREQ}

DESCRIPTION	63	125	250	500	1K	2K	4K	8K
FAN SOUND POWER LEVELS, dB	97	97	87	82	80	75	70	65

COMPUTER PROGRAMS FOR CHAPTER 3

CHILLERS

PROGRAM OUTLINE

Choose reciprocating or centrifugal chiller
Input nominal size of chiller, tons
Calculate 1/1 octave band sound pressure levels
Print out results

COMPUTER PROGRAM – CHILLER.BAS

```
10  '           >EQUIPMENT SOUND POWER: CHILLERS<
20  '
30  DIM T%(8)
40  '
50  '           List of Variables
60  '
70  '           FZ = 1/1 Octave Band Center Frequency, Hz
80  '           LP = Overall Sound Pressure Level, dB
90  '           S1 = Selection for Type of Chiller
100 '           TONS = Size of Chiller Nominal Tons
110 '           T%() = 1/1 Octave Band Sound Pressure Level, dB
120 '
130 FOR I = 1 TO 4
140   FOR J = 1 TO 7
150     READ D(I, J)
160   NEXT J
170 NEXT I
180 CLS : PRINT "CHILLER SOUND PRESSURE LEVELS": PRINT
190 PRINT "1. RECIPROCATING CHILLERS"
200 PRINT "2. CENTRIFUGAL CHILLERS"
210 INPUT "SELECTION: ", S1
220 '
230 IF S1 >= 1 AND S1 <= 2 THEN GOTO 250 ELSE GOTO 210
240 '
250 PRINT
260 INPUT "SIZE OF CHILLER (NOMINAL TONS): ", TONS
270 PRINT
280 '
290 PRINT "CONTINUE? (Y OR N)"
300 Q$ = INKEY$: IF Q$ = "Y" OR Q$ = "N" OR Q$ = "y" OR Q$ = "n" THEN GOTO 310 ELSE GOTO 300
310 IF Q$ = "N" OR Q$ = "n" THEN GOTO 180
320 ON S1 GOSUB 420, 510
330 '
340 PRINT SPC(45); "1/1 OCT BAND CENTER FREQ"
350 PRINT SPC(10); "DESCRIPTION"; SPC(21); " 63 125 250 500  1K   2K   4K"
360 PRINT STRING$(41, CHR$(45)) + " ___ ___ ___ ___ ___ ___ ___"
370 PRINT USING "\                                                          \"; "CHILLER SOUND PRESSURE LEVEL, dB";
380 PRINT USING " ###"; T%(1); T%(2); T%(3); T%(4); T%(5); T%(6); T%(7)
390 '
400 END
410 '
420 '       >SUBROUTINE FOR RECIPROCATING CHILLERS<
430 '
440 LP = 71 + 9 * LOG(TONS) / LOG(10): ' equation 3.4
450 FOR I = 1 TO 7
460   FZ = 62.5 * 2 ^ (I - 1)
470   T%(I) = LP + D(4, I)
480 NEXT I
490 RETURN
500 '
510 '       >SUBROUTINE FOR CENTRIFUGAL CHILLERS<
520 '
530 PRINT "CENTRIFUGAL CHILLERS"
540 PRINT "   1.   Internal Geared, Medium to Full Load"
550 PRINT "   2.   Direct Drive, Medium to Full Load"
560 PRINT "   3.   >1000 Ton, Medium to Full Load"
570 INPUT "SELECTION: ", S2
580 '
```

```
590 LP = 60 + 11 * LOG(TONS) / LOG(10): ' equation 3.3
600 '
610 FOR I = 1 TO 7
620     FZ = 62.5 * 2 ^ (I - 1)
630     T%(I) = LP + D(S2, I)
640 NEXT I
650 RETURN
660 '
670 DATA -8,-5,-6,-7,-8,-5,-8: ' Centrifugal Chiller, Internal Geared, Medium to Full Load"
680 DATA -8,-6,-7,-3,-4,-7,-12: ' Centrifugal Chiller, Direct Drive, Medium to Full Load"
690 DATA -11,-11,-8,-8,-4,-6,-13: ' Centrifugal Chiller, >1000 Ton, Medium to Full Load"
700 DATA -19,-11,-7,-1,-4,-9,-14: ' Reciprocating Chiller, All Loads
```

SAMPLE OUTPUT

1. Reciprocating Chillers
2. Centrifugal Chillers
SELECTION: 1

SIZE OF CHILLER (Nominal Tons): 100

CONTINUE? (Y OR N)

DESCRIPTION	\multicolumn{7}{c}{1/1 OCT BAND CENTER FREQ}
	63	125	250	500	1K	2K	4K
CHILLER SOUND PRESSURE LEVEL, dB	70	78	82	88	85	80	75

COMPUTER PROGRAMS FOR CHAPTER 4

APPENDIX 3

COMPUTER PROGRAMS FOR CHAPTER 4

DAMPERS

PROGRAM OUTLINE

Input system parameters:
- Duct dimensions
- Pressure loss across damper, in H$_2$O
- Duct volume flow rate, cfm
- Duct height normal to damper axis, in
- Type of damper

Calculate:
- Cross–sectional area, ft^2
- Total pressure loss coefficient, C
- Blockage factor, BF
- Flow velocity in damper constriction, ft/s
- Strouhal number, S$_t$
- Sound power level, dB

COMPUTER PROGRAM – DAMPWL.BAS

```
10  '>DAMPER SOUND POWER COMPUTER PROGRAM<
20  '
30  DIM T%(8)
40  PI = 3.141593
50  '
60  '     LIST OF VARIABLES
70  '
80  '     AR = Cross sectional area of duct, sq. ft.
90  '     BF = Blockage factor
100 '     CE = Total pressure loss coefficient
110 '     DD = Duct diameter, in
120 '     DH = Larger rectangular dimension, in
130 '     DN = Duct height normal to damper axis, in
140 '     DP = Pressure loss across damper, in H20
150 '     DW = Smaller rectangular dimension, in
160 '     FZ = 1/1 octave band center frequency, Hz
170 '     KDE = Characteristic spectrum, dB
180 '     PW = For use in calculating sound power level (eq. 4.1)
190 '     ST = Strouhal number
200 '     S1 = Selection for rectangular or circular ducts
210 '     S2 = Selection for type of damper
220 '     T%() = Output sound power level, dB (eq. 4.1)
230 '     VC = Flow velocity in the damper constriction, ft/sec
240 '     VF = Duct volume flow rate, cfm
250 '
260 CLS : PRINT "REGENERATED SOUND POWER LEVELS OF A DAMPER IN A DUCT": PRINT
270 PRINT "1. RECTANGULAR DUCT         2. CIRCULAR DUCT"
280 INPUT "SELECTION: ", S1: PRINT
290 IF S1 >= 1 AND S1 <= 2 THEN GOTO 300 ELSE GOTO 280
300 IF S1 = 2 THEN GOTO 400
310 '
320 ' Input statements for rectangular ducts
330 '
340 INPUT "ENTER SMALLER DUCT DIMENSION (IN): ", DW
350 INPUT "ENTER LARGER DUCT DIMENSION (IN): ", DH
360 DD = 0: GOTO 420
370 '
380 ' Input statement for circular ducts
390 '
400 INPUT "ENTER DUCT DIAMETER (IN): ", DD
410 DW = 0: DH = 0
420 INPUT "ENTER PRESSURE LOSS ACROSS DAMPER (IN H20): ", DP
430 INPUT "ENTER DUCT VOLUME FLOW (CFM): ", VF
440 INPUT "ENTER DUCT HEIGHT NORMAL TO DAMPER AXIS (IN): ", DN: DN = DN / 12
450 PRINT " ": PRINT "DAMPER TYPE": PRINT "1. SINGLE PLATE"; SPC(10); "2. MULTI-VANE"
```

```
460 INPUT "SELECTION: ", S2
470 IF S2 >= 1 AND S2 <= 2 THEN GOTO 480 ELSE GOTO 460
480 PRINT
490 PRINT "CONTINUE? (Y OR N)"
500 Q$ = INKEY$: IF Q$ = "Y" OR Q$ = "N" OR Q$ = "y" OR Q$ = "n" THEN GOTO 510 ELSE GOTO 500
510 IF Q$ = "N" OR Q$ = "n" THEN GOTO 260
520 IF S1 = 1 THEN AR = DW * DH / 144: ' For rectangualr ducts (sq. ft.)
530 IF S1 = 2 THEN AR = PI * DD ^ 2 / 576: ' For circular ducts (sq. ft.)
540 CE = 1.59E+07 * DP * AR ^ 2 / VF ^ 2: ' Equation 4.2
550 IF CE = 1 THEN CE = .999: ' Eliminates division by zero problem
560 IF S2 = 2 THEN BF = (SQR(CE) - 1) / (CE - 1): ' For multi-blade dampers (4.3a)
570 IF S2 = 1 AND CE <= 4 THEN BF = (SQR(CE) - 1) / (CE - 1): ' For single-blade dampers (4.3b)
580 IF S2 = 1 AND CE > 4 THEN BF = .68 * CE ^ (-.15) - .22: ' For single-blade dampers (4.3b)
590 VC = .0167 * VF / (AR * BF): ' Equation 4.4
600 '
610 ' Equation 4.1 without first two terms
620 '
630 PW = 50 * LOG(VC) / LOG(10) + 10 * LOG(AR) / LOG(10) + 10 * LOG(DN) / LOG(10)
640 '
650 FOR I = 1 TO 8
660 FZ = 62.5 * 2 ^ (I - 1): ST = FZ * DN / VC: ' Equation 5
670 IF ST <= 25 THEN KDE = -36.3 - 10.7 * LOG(ST) / LOG(10): ' equation 4.6
680 IF ST > 25 THEN KDE = -1.1 - 35.9 * LOG(ST) / LOG(10): ' equation 4.6
690 T%(I) = PW + KDE + 10 * LOG(FZ / 62.5) / LOG(10): ' equation 4.1 inclusive
700 NEXT I
710 '
720 ' Output statements
730 '
740 PRINT
750 PRINT SPC(45); "1/1 OCT BAND CENTER FREQ"
760 PRINT SPC(10); "DESCRIPTION"; SPC(21); " 63 125 250 500  1K  2K  4K  8K"
770 PRINT STRING$(41, CHR$(45)) + " ___ ___ ___ ___ ___ ___ ___ ___"
780 PRINT USING "\                                                            \"; "REGENERATED SOUND POWER LEVELS, dB";
790 PRINT USING " ###"; T%(1); T%(2); T%(3); T%(4); T%(5); T%(6); T%(7); T%(8)
800 END
```

SAMPLE OUTPUT FOR EXAMPLE 4.1

REGENERATED SOUND POWER LEVELS OF A DAMPER IN A DUCT

1. RECTANGULAR DUCT   2. CIRCULAR DUCT
SELECTION: 1

ENTER SMALLER DUCT DIMENSION (IN): 12
ENTER LARGER DUCT DIMENSION (IN): 12
ENTER PRESSURE LOSS ACROSS DAMPER (IN H20): .5
ENTER DUCT VOLUME FLOW (CFM): 4000
ENTER DUCT HEIGHT NORMAL TO DAMPER AXIS (IN): 12

DAMPER TYPE
1. SINGLE PLATE   2. MULTI-VANE
SELECTION: 2

CONTINUE (Y OR N)

			1/1 OCT BAND CENTER FREQ					
DESCRIPTION	63	125	250	500	1K	2K	4K	8K
REGENERATED SOUND POWER LEVELS, dB	69	69	69	69	68	68	64	57

ELBOWS FITTED WITH TURNING VANES

PROGRAM OUTLINE

Input system parameters:
    Duct dimensions
    Pressure loss across bend
    Duct volume flow rate
    Duct height normal to vane length
    Cord length of vanes
    Number of vanes

Calculate:
    Total pressure loss coefficient, C
    Blockage factor, BF
    Constricted flow velocity, U_c
    Strouhal number, S_t
    Characteristic spectrum, K_T
    Sound power level

COMPUTER PROGRAM – BENDVANE.BAS

```
10 '>SOUND POWER LEVEL OF BENDS FITTED WITH TURNING VANES COMPUTER PROGRAM<
20 '
30 DIM T%(8)
40 '
50 '      LIST OF VARIABLES
60 '
70 '      AR = cross-sectional area of duct, sq. ft.
80 '      BF = blockage factor (Eq. 3)
90 '      CE = total pressure loss coefficient
100 '     CL = cord length of vanes
110 '     DA = duct height normal to vane length, in
120 '     DH = larger duct dimension, in
130 '     DN = duct height normal to vane length, ft
140 '     DP = pressure loss across bend, in H20
150 '     DW = smaller duct dimension, in
160 '     FZ = 1/1 octave band center frequency, Hz
170 '     KTE = characteristic spectrum
180 '     NV = number of vanes
190 '     PW = constant terms in eq. 4.7 for sound power level
200 '     ST = Strouhal number
210 '     VC = flow velocity in the turning vane constriction, fps
220 '     VF = duct volume flow rate, cfm
230 '     T%() = sound power level output, dB
240 CLS : PRINT "REGENERATED SOUND POWER LEVELS OF": PRINT "BENDS FITTED WITH TURNING VANES IN RECTANGULAR DUCTS"
260 PRINT
270 INPUT "ENTER SMALLER DUCT DIMENSION (IN): ", DW
280 INPUT "ENTER LARGER DUCT DIMENSION (IN): ", DH
290 INPUT "ENTER PRESSURE LOSS ACROSS BEND (IN H20): ", DP
300 INPUT "ENTER DUCT VOLUME FLOW (CFM): ", VF
310 INPUT "ENTER DUCT HEIGHT NORMAL TO VANE LENGTH (IN): ", DA
320 INPUT "ENTER CORD LENGTH OF VANES (IN): ", CL
330 INPUT "ENTER NUMBER OF VANES: ", NV
340 PRINT
350 PRINT "CONTINUE? (Y OR N)"
360 Q$ = INKEY$: IF Q$ = "Y" OR Q$ = "N" OR Q$ = "y" OR Q$ = "n" THEN GOTO 370 ELSE GOTO 360
370 IF Q$ = "N" OR Q$ = "n" THEN GOTO 250
380 AR = DW * DH / 144: DN = DA / 12
390 CE = 1.59E+07 * DP * AR ^ 2 / VF ^ 2: ' Equation 4.8 (English units)
400 IF CE = 1 THEN CE = .999: ' Eliminates division by zero problem
410 BF = (SQR(CE) - 1) / (CE - 1): ' Equation 4.9
420 VC = .0167 * VF / (AR * BF): ' Equation 4.10
430 '
```

```
440 ' Equation 4.7 without the first two terms
450 PW = 50 * LOG(VC) / LOG(10) + 10 * LOG(AR) / LOG(10) + 10 * LOG(CL) / LOG(10) + 10 * LOG(NV) / LOG(10)
460 '
470 FOR I = 1 TO 8
480 FZ = 62.5 * 2 ^ (I - 1): ST = FZ * DN / VC: ' Equation 4.11
490 KTE = -47.5 - 7.69 * ABS(LOG(ST) / LOG(10)) ^ 2.5: ' Equation 4.12
500 T%(I) = PW + KTE + 10 * LOG(FZ / 62.5) / LOG(10): ' Equation 4.7 (inclusive)
510 IF T%(I) < 0 THEN T%(I) = 0
520 NEXT I
530 '
540 ' Output statements
550 '
560 PRINT
570 PRINT SPC(45); "1/1 OCT BAND CENTER FREQ"
580 PRINT SPC(10); "DESCRIPTION"; SPC(21); " 63  125  250  500   1K   2K   4K   8K"
590 PRINT STRING$(41, CHR$(45)) + "  ___ ___ ___ ___ ___ ___ ___ ___"
600 PRINT USING "\                                        \"; "REGENERATED SOUND POWER LEVELS, dB";
610 PRINT USING "  ###"; T%(1); T%(2); T%(3); T%(4); T%(5); T%(6); T%(7); T%(8)
620 END
```

SAMPLE OUTPUT FOR EXAMPLE 4.2

REGENERATED SOUND POWER LEVELS OF
ELBOWS FITTED WITH TURNING VANES IN RECTANGULAR DUCTS

ENTER SMALLER DUCT DIMENSION (IN): 20
ENTER LARGER DUCT DIMENSION (IN): 20
ENTER PRESSURE LOSS ACROSS BEND (IN H20): .16
ENTER DUCT VOLUME FLOW (CFM): 8500
ENTER DUCT HEIGHT NORMAL TO VANE LENGTH (IN): 20
ENTER CORD LENGTH OF VANES (IN): 7.9
ENTER NUMBER OF VANES: 5

CONTINUE? (Y OR N)

DESCRIPTION	63	125	250	500	1K	2K	4K	8K
REGENERATED SOUND POWER LEVELS, dB	67	70	70	68	64	56	46	31

1/1 OCT BAND CENTER FREQ

JUNCTIONS AND TURNS

PROGRAM OUTLINE

Input system parameters:
  Type of duct junction
  Geometry of feeder duct
  Dimensions of feeder duct
  Feeder duct volume flow rate, cfm
  Geometry of branch duct
  Dimensions of branch duct
  Branch duct volume flow rate, cfm
  Radius or turn of junction or branch duct, in
  Flow condition; turbulent or laminar

Calculate:
  Cross-sectional areas
  Flow velocities
  Rounding parameter
  Strouhal number
  Δr
  ΔT, correction factor for upstream turbulence
  K_J, characteristic spectrum
  $L_W(f_o)_b$, branch sound power levels
  $L_W(f_o)_m$, main duct sound power levels

COMPUTER PROGRAM – JUNCT.BAS

```
10 REM >REGENERATED SOUND POWER OF DUCT JUNCTIONS<
20 DIM T%(8)
30 PI = 3.141593
40 CLS
50 '
60 '     LIST OF VARIABLES
70 '
80 '     AB  = cross-sectional area of branch duct, sq. ft.
90 '     AM  = corss-sectional area of main duct, sq. ft.
100 '    AR  = area ratio
110 '    BF  = branch duct volume flow rate, cfm
120 '    BV  = flow velocity in branch duct, fps
130 '    CD  = equivalent diameter, ft
140 '    DB  = equivalent diameter of branch duct, ft
150 '    DDB = branch duct diameter, in
160 '    DDM = main duct diameter, in
170 '    DHB = larger branch duct dimension, in
180 '    DHM = larger rectangular main duct dimension, in
190 '    DM  = euivalent diameter of main duct, ft
200 '    DR  = correction term for size of radius of the bend
210 '    DT  = correction for upstream turbulence
220 '    DWB = smaller rectangular branch duct dimension, in
230 '    DWM = smaller rectangular main duct dimension, in
240 '    FZ  = 1/1 octave band center frequency, Hz
250 '    KJE = characteristic spectrum
260 '    MF  = main duct volume flow rate, cfm
270 '    MV  = flow velocity in main duct, fps
280 '    PW  = constant terms of equation 4.13 for sound power level
290 '    RA  = radius of turn or junction of branch duct, in
300 '    RD  = rounding parameter
310 '    ST  = Strouhal number
320 '    S1  = selection for type of junction
330 '    S2  = selection for geometry of main feeder duct
340 '    S3  = selection for geometry of branch duct
350 '    S4  = selection for flow conditions
```

```
360 '       VR = velocity ratio
370 '
380 PRINT "REGENERATED SOUND POWER OF DUCT JUNCTIONS": PRINT
390 PRINT " 1. SOUND POWER: 90 DEGREE BEND W/O TURNING VANES"
400 PRINT " 2. SOUND POWER: DUCT T-JUNCTION"
410 PRINT " 3. SOUND POWER: DUCT X-JUNCTION"
420 PRINT " 4. SOUND POWER: DUCT 90 DEGREE BRANCH TAKEOFF"
430 INPUT "SELECTION: ", S1
440 IF S1 >= 1 AND S1 <= 4 THEN GOTO 450 ELSE GOTO 430
450 PRINT
460 IF S1 = 1 THEN PRINT "REGENERATED SOUND POWER LEVELS OF A 90 DEG BEND W/O TURNING VANES"
470 IF S1 = 2 THEN PRINT "REGENERATED SOUND POWER LEVELS OF A DUCT T-JUNCTION"
480 IF S1 = 3 THEN PRINT "REGENERATED SOUND POWER LEVELS OF A DUCT X-JUNCTION"
490 IF S1 = 4 THEN PRINT "REGENERATED SOUND POWER LEVELS OF A DUCT 90 DEG BRANCH TAKEOFF"
500 '
510 PRINT "MAIN FEEDER DUCT:   (1. RECTANGULAR       2. CIRCULAR)"
520 INPUT "SELECTION: ", S2
530 IF S2 >= 1 AND S2 <= 2 THEN GOTO 540 ELSE GOTO 520
540 IF S2 = 2 THEN GOTO 600
550 '
560 INPUT "ENTER SMALLER DUCT DIMENSION (IN): ", DWM
570 INPUT "ENTER LARGER DUCT DIMENSION (IN): ", DHM
580 DDM = 0: GOTO 630
590 '
600 INPUT "ENTER DUCT DIAMETER (IN): ", DDM
610 DWM = 0: DHM = 0
620 '
630 INPUT "ENTER DUCT VOLUME FLOW (CFM): ", MF
640 '
650 IF S1 = 1 THEN S3 = 0: DWB = 0: DHB = 0: DDB = 0: BF = 0: GOTO 800
660 '
670 PRINT : PRINT "BRANCH DUCT:   (1. RECTANGULAR       2. CIRCULAR)"
680 INPUT "SELECTION: ", S3
690 IF S3 >= 1 AND S3 <= 2 THEN GOTO 700 ELSE GOTO 680
700 '
710 IF S3 = 2 THEN GOTO 770: ' S3 = 2 when branch duct is circular
720 '
730 INPUT "ENTER SMALLER DUCT DIMENSION (IN): ", DWB
740 INPUT "ENTER LARGER DUCT DIMENSION (IN): ", DHB
750 DDB = 0: GOTO 790
760 '
770 INPUT "ENTER DUCT DIAMETER (IN): ", DDB
780 DWB = 0: DHB = 0
790 INPUT "ENTER DUCT VOLUME FLOW (CFM): ", BF
800 INPUT "ENTER RADIUS OF TURN OR JUNCTION OF BRANCH DUCT (IN.): ", RA
810 '
820 IF S2 = 1 THEN AM = DWM * DHM / 144: MV = MF / (60 * AM)
830 IF S2 = 2 THEN AM = PI * DDM ^ 2 / 576: MV = MF / (60 * AM)
840 '
850 PRINT : PRINT "ARE THERE DAMPERS, TURNS, TAKEOFFS, ETC. WITHIN A DISTANCE OF": PRINT INT(50 * SQR(4 * AM / PI + .5)) / 10; " FT OF THE JUNCTION?(1. YES   2. NO)"
860 INPUT "SELECTION: ", S4
870 IF S4 >= 1 AND S4 <= 2 THEN GOTO 880 ELSE GOTO 860
880 PRINT
890 PRINT "CONTINUE? (Y OR N)"
900 Q$ = INKEY$: IF Q$ = "Y" OR Q$ = "N" OR Q$ = "y" OR Q$ = "n" THEN GOTO 910 ELSE GOTO 900
910 IF Q$ = "N" OR Q$ = "n" THEN GOTO 390
920 IF S1 = 1 THEN S4 = 2: GOTO 940
930 ' CD = equivalent diameter, AB = area of branch, BV = branch velocity
940 IF S1 = 1 THEN BV = MV: AB = AM: CD = SQR(4 * AM / PI): GOTO 990: ' 90 degree bend w/o turning vanes
950 IF S3 = 1 THEN AB = DWB * DHB / 144: CD = SQR(4 * AB / PI): BV = BF / (60 * AB): ' rectangular branch
960 IF S3 = 2 THEN AB = PI * DDB ^ 2 / 576: CD = DDB / 12: BV = BF / (60 * AB): ' circular branch
970 '
980 DB = SQR(4 * AB / PI): DM = SQR(4 * AM / PI): ' equivalent diameters for branch and main ducts
990 RD = RA / (12 * CD): VR = MV / BV: AR = AM / AB
1000 IF S4 = 2 THEN DT = 0
1010 IF S4 = 1 THEN DT = -1.667 + 1.8 * VR - .133 * VR ^ 2: ' Equation 4.20
1020 PW = 50 * LOG(BV) / LOG(10) + 10 * LOG(AB * CD) / LOG(10) + DT: ' Part of Eq. 4.14
1030 '
1040 FOR I = 1 TO 8
1050 FZ = 62.5 * 2 ^ (I - 1): ST = CD * FZ / BV: ' Frequency and Equation 4.19
1060 '   Characteristic specturm (Eq. 4.22)
```

```
1070 KJE = -21.6 + 12.388 * VR ^ .673 - 16.482 * VR ^ (-.303) * LOG(ST) / LOG(10) - 5.047 * VR ^
     (-.254) * (LOG(ST) / LOG(10)) ^ 2
1080 DR = (1 - RD / .15) * (6.793 - 1.86 * LOG(ST) / LOG(10)): ' Equation 4.17
1090 T%(I) = PW + KJE + DR + 10 * LOG(FZ / 62.5) / LOG(10): ' Branch Pwl (Eq. 4.13)
1100 IF T%(I) <= 0 THEN T%(I) = 0: ' Eliminates negative sound power calculations
1110 NEXT I
1120 '
1130 ' Output statments
1140 '
1150 PRINT
1160 PRINT SPC(45); "1/1 OCT BAND CENTER FREQ"
1170 PRINT SPC(10); "DESCRIPTION"; SPC(21); " 63  125  250  500   1K   2K   4K   8K"
1180 A$ = STRING$(41, CHR$(45)) + "  ___  ___  ___  ___  ___  ___  ___  ___"
1190 PRINT A$
1200 PRINT USING "\                                                    \"; "BRANCH REGENERATED SOUND POWER
     LEVELS, dB";
1210 PRINT USING "### "; T%(1); T%(2); T%(3); T%(4); T%(5); T%(6); T%(7); T%(8)
1220 ' Main duct sound power levels, dB
1230 FOR I = 1 TO 8
1240 IF T%(I) = 0 THEN GOTO 1280
1250 IF S1 = 2 THEN T%(I) = T%(I) + 3: ' For a Duct T-Junction
1260 IF S1 = 3 THEN T%(I) = T%(I) + 20 * LOG(DM / DB) / LOG(10) + 3: ' For a Duct X-Junction
1270 IF S1 = 4 THEN T%(I) = T%(I) + 20 * LOG(DM / DB) / LOG(10): ' For a 90 deg. Branch takeoff
1280 NEXT I
1290 PRINT
1300 PRINT USING "\                                                    \"; "MAIN REGENERATED SOUND POWER
     LEVELS, dB";
1310 PRINT USING "### "; T%(1); T%(2); T%(3); T%(4); T%(5); T%(6); T%(7); T%(8)
```

SAMPLE OUTPUT FOR EXAMPLE 4.3: X-Junction

```
  1. SOUND POWER: 90 DEGREE BEND W/O TURNING VANES
  2. SOUND POWER: DUCT T-JUNCTION
  3. SOUND POWER: DUCT X-JUNCTION
  4. SOUND POWER: DUCT 90 DEGREE BRANCH TAKEOFF
SELECTION: 3

REGENERATED SOUND POWER LEVELS OF DUCT X-JUNCTION
MAIN FEEDER DUCT:  (1. RECTANGULAR       2. CIRCULAR)
SELECTION: 1
ENTER SMALLER DUCT DIMENSION (IN): 12
ENTER LARGER DUCT DIMENSION (IN): 36
ENTER DUCT VOLUME FLOW (CFM): 12000

BRANCH DUCT:  (1. RECTANGULAR       2. CIRCULAR)
SELECTION: 1
ENTER SMALLER DUCT DIMENSION (IN): 10
ENTER LARGER DUCT DIMENSION (IN): 10
ENTER DUCT VOLUME FLOW (CFM): 1200
ENTER RADIUS OF TURN OR JUNCTION OF BRANCH DUCT (IN.): 0

ARE THERE DAMPERS, TURNS, TAKEOFFS, ETC. WITHIN A DISTANCE OF
 10.3  FT OF THE JUNCTION?(1. YES     2. NO)
SELECTION: 2

CONTINUE? (Y OR N)
```

DESCRIPTION	63	125	250	500	1K	2K	4K	8K
BRANCH REG. SOUND POWER LEVELS, dB	73	71	67	63	59	53	47	40
MAIN REG. SOUND POWER LEVELS, dB	82	80	76	72	68	62	56	49

SAMPLE OUTPUT FOR EXAMPLE 4.4: T–Junction

```
1. SOUND POWER: 90 DEGREE BEND W/O TURNING VANES
2. SOUND POWER: DUCT T-JUNCTION
3. SOUND POWER: DUCT X-JUNCTION
4. SOUND POWER: DUCT 90 DEGREE BRANCH TAKEOFF
SELECTION: 2

REGENERATED SOUND POWER LEVELS OF A DUCT T-JUCNTION
MAIN FEEDER DUCT: (1. RECTANGULAR    2. CIRCULAR)
SELECTION: 1
ENTER SMALLER DUCT DIMENSION (IN): 12
ENTER LARGER DUCT DIMENSION (IN): 36
ENTER DUCT VOLUME FLOW (CFM): 12000

BRANCH DUCT: (1. RECTANGULAR    2. CIRCULAR)
SELECTION: 1
ENTER SMALLER DUCT DIMENSION (IN): 12
ENTER LARGER DUCT DIMENSION (IN): 18
ENTER DUCT VOLUME FLOW (CFM): 6000
ENTER RADIUS OF TURN OR JUNCTION OF BRANCH DUCT (IN.): 0

ARE THERE DAMPERS, TURNS, TAKEOFFS, ETC. WITHIN A DISTANCE OF
 10.3 FT OF THE JUNCTION?(1. YES    2. NO)
SELECTION: 2

CONTINUE? (Y OR N)
```

DESCRIPTION	63	125	250	500	1K	2K	4K	8K
BRANCH REG. SOUND POWER LEVELS, dB	90	87	82	77	71	64	56	47
MAIN REG. SOUND POWER LEVELS, dB	93	90	85	80	74	67	59	50

SAMPLE OUTPUT FOR EXAMPLE 4.5: 90° Elbow without Turning Vanes

```
1. SOUND POWER: 90 DEGREE BEND W/O TURNING VANES
2. SOUND POWER: DUCT T-JUNCTION
3. SOUND POWER: DUCT X-JUNCTION
4. SOUND POWER: DUCT 90 DEGREE BRANCH TAKEOFF
SELECTION: 1

REGENERATED SOUND POWER LEVELS OF A 90 DEG BEND W/O TURNING VANES
MAIN FEEDER DUCT: (1. RECTANGULAR    2. CIRCULAR)
SELECTION: 1
ENTER SMALLER DUCT DIMENSION (IN): 12
ENTER LARGER DUCT DIMENSION (IN): 36
ENTER DUCT VOLUME FLOW (CFM): 12000
ENTER RADIUS OF TURN OR JUNCTION OF BRANCH DUCT (IN.): 0

ARE THERE DAMPERS, TURNS, TAKEOFFS, ETC. WITHIN A DISTANCE OF
 10.3 FT OF THE JUNCTION?(1. YES    2. NO)
SELECTION: 2

CONTINUE? (Y OR N)
```

DESCRIPTION	63	125	250	500	1K	2K	4K	8K
BRANCH REG. SOUND POWER LEVELS, dB	91	88	83	77	71	63	55	46
MAIN REG. SOUND POWER LEVELS, dB	91	88	83	77	71	63	55	46

SAMPLE OUTPUT FOR EXAMPLE 4.6: 90° Branch Takeoff

```
1. SOUND POWER: 90 DEGREE BEND W/O TURNING VANES
2. SOUND POWER: DUCT T-JUNCTION
3. SOUND POWER: DUCT X-JUNCTION
4. SOUND POWER: DUCT 90 DEGREE BRANCH TAKEOFF
SELECTION: 3

REGENERATED SOUND POWER LEVELS OF DUCT 90 DEG BRANCH TAKEOFF
MAIN FEEDER DUCT: (1. RECTANGULAR    2. CIRCULAR)
SELECTION: 1
ENTER SMALLER DUCT DIMENSION (IN): 12
ENTER LARGER DUCT DIMENSION (IN): 36
ENTER DUCT VOLUME FLOW (CFM): 12000

BRANCH DUCT: (1. RECTANGULAR    2. CIRCULAR)
SELECTION: 1
ENTER SMALLER DUCT DIMENSION (IN): 10
ENTER LARGER DUCT DIMENSION (IN): 10
ENTER DUCT VOLUME FLOW (CFM): 1200
ENTER RADIUS OF TURN OR JUNCTION OF BRANCH DUCT (IN.): 0

ARE THERE DAMPERS, TURNS, TAKEOFFS, ETC. WITHIN A DISTANCE OF
 10.3 FT OF THE JUNCTION?(1. YES    2. NO)
SELECTION: 2

CONTINUE? (Y OR N)
```

DESCRIPTION	63	125	250	500	1K	2K	4K	8K
BRANCH REG. SOUND POWER LEVELS, dB	73	71	67	63	59	53	47	40
MAIN REG. SOUND POWER LEVELS, dB	79	77	73	69	65	59	53	46

(1/1 OCT BAND CENTER FREQ)

DIFFUSERS

PROGRAM OUTLINE

Input system parameters:
    Type of diffuser
    Duct flow volume, cfm
    Pressure drop across diffuser, in H_2O
    Duct or diffuser dimensions, in

Calculate:
    Duct or diffuser cross-section area, ft^2
    Flow velocity, ft/sec
    Pressure drop coefficient
    Sound power level, dB

COMPUTER PROGRAM – DIFFUSER.BAS

```
10  '     >AIR DIFFUSER REGENERATED SOUND POWER
20  '
30  DIM T%(8), NC%(11, 7), U%(7)
40  PI = 3.1415927#: ROW = .0749
50  '
60  '       LIST OF VARIABLES
70  '
80  '       NCC = NC SPECIFIED OR NOT
90  '       NCV = NC SPECIFIED VALUE
100 '       DT  = DIFFUSER TYPE
110 '       FR  = FLOW RATE
120 '       PD  = PRESSURE DROP
130 '       DL  = DIFFUSER LENGTH
140 '       DW  = DIFFUSER WIDTH
150 '       DD  = DIFFUSER DIAMETER
160 '       S   = CROSS SECTIONAL AREA
170 '       U   = MEAN FLOW SPEED
180 '       E   = NORMALIZED PRESSURE DROP COEFFICIENT
190 '       LG  = TO CONVERT NATURAL LOG TO LOG BASE 10
200 '       LW  = SOUND POWER LEVEL
210 '       F   = PEAK FREQUENCY
220 '       CO  = CORRECTION VALUE TO BE ADDED TO LW
230 '       T%()= CORRECTED SOUND POWER LEVELS
240 '       ROW = .0749 pcf THE DENSITY OF AIR
250 '
260 CLS : PRINT "AIR DIFFUSER REGENERATED SOUND POWER": PRINT
270 PRINT "DETERMINE DIFFUSER SOUND POWER LEVELS BY MEANS OF:"
280 PRINT "1. ENTERING MANUFACTURER'S DIFFUSSER NC-RATING"
290 PRINT "2. ENTERING GENERIC DIFFUSER TYPE AND DIMENSIONS"
300 INPUT "SELECTION: ", NCC
310 PRINT
320 IF NCC = 1 THEN GOSUB 350: GOSUB 500: GOTO 600
330 IF NCC = 2 THEN GOSUB 440: GOSUB 500: GOTO 600
340 IF NCC > 2 OR NCC < 1 THEN NCC = 0: LOCATE 6, 62: PRINT "            ": GOTO 300
350 PRINT "THE NC-VALUE MUST BE BETWEEN 15 AND 65."
360 PRINT "ENTER THE SPECIFIED NC-LEVEL."
370 INPUT "NC = ", NCV
380 IF NCV < 15 OR NCV > 65 THEN GOTO 370
390 PRINT
400 PRINT "1. RECTANGULAR DIFFUSER           2. ROUND DIFFUSER"
410 INPUT "SELECTION: ", DT
420 RETURN
430 PRINT "GENERIC DIFFUSER TYPE"
440 PRINT "1. RECTANGULAR DIFFUSERS"
450 PRINT "2. ROUND DIFFUSERS"
460 PRINT "3. SQUARE PERFORATED FACE DIFFUSERS (WITH ROUND INLET)"
470 PRINT "4. RECTANGULAR SLOT DIFFUSERS"
480 INPUT "ENTER THE DIFFUSER TYPE: ", DT
490 RETURN
500 PRINT : INPUT "ENTER THE FLOW RATE (CFM): ", FR
```

```
510  IF NCC = 1 THEN GOTO 530
520  INPUT "ENTER THE PRESSURE DROP (IN H2O): ", PD
530  IF DT <> 2 THEN GOSUB 550: GOTO 600
540  IF DT = 2 THEN GOSUB 580: GOTO 600
550  PRINT : INPUT "ENTER THE LENGTH OF THE DIFFUSER (IN): ", DL
560  INPUT "ENTER THE WIDTH OF THE DIFFUSER (IN): ", DW
570  RETURN
580  PRINT : INPUT "ENTER THE DIAMETER OF THE DIFFUSER (IN): ", DD
590  RETURN
600  PRINT : PRINT "CONTINUE (Y/N) "
610  Q$ = INKEY$: IF Q$ = "N" OR Q$ = "Y" OR Q$ = "y" OR Q$ = "n" THEN GOTO 620 ELSE GOTO 610
620  IF Q$ = "N" OR Q$ = "n" THEN GOTO 260
630  IF DT = 2 THEN ZZ = 0 ELSE ZZ = -6
640  IF DT = 2 OR DT = 3 THEN S = PI * (DD / 2) ^ 2 / 144
650  IF DT = 1 OR DT = 4 THEN S = DL * DW / 144
660  U = FR / (60 * S):    ' Equation 4.28
670  E = 334.9 * PD / (ROW * U ^ 2): ' Equation 4.27
680  IF NCC = 1 THEN LW = 10
690  IF NCC = 2 THEN LW = 10 * LOG(S) / LOG(10) + 30 * LOG(E) / LOG(10) + 60 * LOG(U) / LOG(10) - 31.3: ' equation 4.39
700  F = 48.77 * U:  ' Equation 4.30
710  GOSUB 880 'returns II value
720  FOR I = 1 TO 8
730  A = I - II
740  CO = -5.82 - .15 * A - 1.13 * A ^ 2 + ZZ:  ' Equation 4.32
750  T%(I) = LW + CO 'corrected sound power levels - output
760  NEXT I
770  GOSUB 990
780  IF NCV <> 0 THEN GOSUB 1090
790  IF NCV <> 0 THEN GOSUB 1280 ELSE NC$ = "    "
800  T1$ = "DIFFUSER SOUND POWER, dB"
810  PRINT
820  PRINT SPC(48); "1/1 OCT BAND CENTER FREQ"
830  PRINT SPC(10); "DESCRIPTION"; SPC(24); " 63  125  250  500   1K   2K   4K   8K"
840  PRINT STRING$(41, CHR$(45)) + "--- --- --- --- --- --- --- ---"
850  PRINT USING "                                              \                            \"; T1$;
860  PRINT USING " ###"; T%(1); T%(2); T%(3); T%(4); T%(5); T%(6); T%(7); T%(8)
870  END
880     IF F < 44 THEN II = 0
890     IF F >= 44 AND F < 88 THEN II = 1
900     IF F >= 88 AND F < 177 THEN II = 2
910     IF F >= 177 AND F < 355 THEN II = 3
920     IF F >= 355 AND F < 700 THEN II = 4
930     IF F >= 700 AND F < 1420 THEN II = 5
940     IF F >= 1420 AND F < 2840 THEN II = 6
950     IF F >= 2840 AND F < 5680 THEN II = 7
960     IF F >= 5680 AND F < 11360 THEN II = 8
970     IF F > 5680 THEN II = 0
980     RETURN
990     FOR I = 1 TO 11
1000       FOR J = 1 TO 7
1010       READ NC%(I, J)
1020       NEXT J
1030    NEXT I
1040    DATA 80,75,71,68,66,64,63,77,71,67,63,61,59,58,74,67,62,58,56,54,53
1050    DATA 71,64,58,54,51,49,48,67,60,54,49,46,44,43,64,56,50,45,41,39,38
1060    DATA 60,52,45,40,36,34,33,57,48,41,35,31,29,28,54,44,37,31,27,24,22
1070    DATA 51,40,33,26,22,19,17,47,36,29,22,17,14,12
1080    RETURN
1090    FOR I = 1 TO 7: U%(I) = T%(I): NEXT I
1100    MAX = 0: FOR I = 1 TO 7: R(I) = 0: NEXT I
1110    FOR I = 1 TO 7
1120    IF U%(I) > NC%(1, I) THEN NC$ = "NC: NC>65": GOTO 1270
1130    NEXT I
1140    FOR J = 1 TO 7
1150    FOR I = 1 TO 11
1160    IF U%(J) = NC%(I, J) THEN R(J) = 70 - 5 * I: GOTO 1210
1170    IF U%(J) < NC%(I, J) THEN GOTO 1200
1180    IF U%(J) > NC%(I, J) THEN X = (NC%(I - 1, J) - U%(J)) / (NC%(I - 1, J) - NC%(I, J))
1190    R(J) = 70 - 5 * (I - 1) - 5 * X: GOTO 1210
1200    NEXT I
1210    NEXT J
1220    FOR I = 1 TO 7
1230    IF MAX < R(I) THEN MAX = R(I)
1240    NEXT I
```

```
1250        IF MAX < 15 THEN NC$ = "NC: NC < 15": GOTO 1270
1260        MAX = CINT(MAX)
1270        RETURN
1280        NN = 0
1290        NC = CINT(NCV)
1300        IF NC = MAX THEN GOTO 1380
1310        FOR I = 1 TO 8
1320        T%(I) = T%(I) + 1
1330        NEXT I
1340        GOSUB 1090
1350        NN = NN + 1: IF NN = 100 THEN GOTO 1370
1360        IF MAX >= NC - .5 AND MAX <= NC + .5 THEN GOTO 1370 ELSE GOTO 1310
1370        NC$ = "NC: NC=" + STR$(NC)
1380        FOR I = 1 TO 8: T%(I) = T%(I) + 10: NEXT I
1390        RETURN
```

SAMPLE OUTPUT FOR EXAMPLE 4.7

AIR DIFFUSER REGENERATED SOUND POWER

DETERMINE DIFFUSER SOUND POWER LEVELS BY MEANS OF:
1. ENTERING MANUFACTURER'S DIFFUSER NC-RATING
2. ENTERING GENERIC DIFFUSER TYPE AND DIMENSIONS
SELECTION: 2

1. RECTANGULAR DIFFUSERS
2. ROUND DIFFUSERS
3. SQUARE PERFORATED FACE DIFFUSERS (WITH ROUND INLET)
4. RECTANGULAR SLOT DIFFUSERS
ENTER THE DIFFUSER TYPE: 1

ENTER THE FLOW RATE (CFM): 1200
ENTER THE PRESSURE DROP (IN H20): .3

ENTER THE LENGTH OF THE DIFFUSER (IN): 16
ENTER THE WIDTH OF THE DIFFUSER (IN): 12

CONTINUE? (Y OR N)

DESCRIPTION	\multicolumn{8}{c}{1/1 OCT BAND CENTER FREQ}
	63	125	250	500	1K	2K	4K	8K
DIFFUSER SOUND POWER, dB	34	42	48	51	52	51	47	41

APPENDIX 4

COMPUTER PROGRAMS FOR CHAPTER 5

COMPUTER PROGRAMS FOR CHAPTER 5

PLENUM CHAMBERS

PROGRAM OUTLINE

Input data from Table 5.1; Absorption coefficients for building material
Input system parameters:
    Inside height of plenum, ft
    Inside width of plenum, ft
    Inside length of plenum, ft
    Horizontal distance between centers of plenum inlet and outlet, ft
    Vertical distance between centers of plenum inlet and outlet, ft
    Cross–sectional dimensions
        plenum inlet
        plenum outlet
    Sound absorption coefficients of plenum shell
    Sound absorption coefficients of plenum liner
    % Plenum surface covered by plenum liner
    Directivity factor
Calculate:
    Surface areas
    Cut–off frequency, Hz
    σl for the 1/1 octave bands from 63 to 500 Hz
    m, the ratio of the cross–sectional area of the plenum divided the inlet cross–section, (Eq. 5.8)
    Plenum transmission loss, dB

COMPUTER PROGRAM – PLENUM.BAS

```
10  '>PLENUM SOUND TRANSMISSION LOSS COMPUTER PROGRAM<
20  '
30  DIM T%(8), TT(8), D(6, 8)
40  DEFINT I-K
50  PI = 3.141593
60  '
70  '     LIST OF VARIABLES
80  '
90  '     AR = total surface area, sq. ft.
100 '     COF = cut-off frequency, Hz
110 '     D() = sound absorption coefficients
120 '     DI = diameter of plenum opening
130 '     FNATTN63() = attenuation function for 63 Hz octave band
140 '     FNATTN125() = attenuation function for 125 Hz octave band
150 '     FNATTN250() = attenuation function for 250 Hz octave band
160 '     FNATTN500() = attenuation function for 500 Hz octave band
170 '     FR = 1/1 octave band center frequency, Hz
180 '     HI = height of plenum inlet opening
190 '     HO = height of plenum outlet opening
200 '     M = ratio of cross sectional areas, equation 5.8
210 '     OD = diameter of plenum outlet opening
220 '     PA = P/A for use with equations 5.9-5.12
230 '     PC = percentage of surface area covered by sound absorbing material
240 '     PD = horizontal distance between centers of plenum inlet and outlet
250 '     PH = inside height of plenum, ft
260 '     PL = inside length of plenum, ft
270 '     PV = vertical distance between centers of plenum inlet and outlet
280 '     PW = inside width of plenum, ft
290 '     Q = directivity factor
300 '     RR = r; figure 1, equation 5.3
310 '     SI = inlet cross-sectional area, sq. ft.
320 '     SO = outlet cross-sectional area, sq. ft.
330 '     SP = cross sectional area of plenum, sq. ft.
```

```
340 '       S1 = selection for type of inlet opening
350 '       S2 = selection for type of outlet opening
360 '       S3 = selection for non-sound absorbing plenum surface
370 '       S4 = selection for sound absorbing plenum liner
380 '       S5 = selection fro Q
390 '       T  = thickness of fiberglass insulation, in
400 '       TH = theta used in equation 5.4
410 '       TT() = used with equation 5.7
420 '       T%() = sound attenuation (equation 5.1 or 5.7), dB
430 '       WI = width of plenum inlet opening
440 '       WO = width of plenum outlet opening
450 '
460 DEF FNCOSH (X) = (EXP(X) + EXP(-X)) / 2: ' Defines cosh function
470 DEF FNSINH (X) = (EXP(X) - EXP(-X)) / 2: ' Defines sinh function
480 DEF FNATTN63 (PA, T, L) = .0133 * PA ^ 1.959 * T ^ .917 * L: ' Eq. 5.9
490 DEF FNATTN125 (PA, T, L) = .0574 * PA ^ 1.41 * T ^ .941 * L: ' Eq. 5.10
500 DEF FNATTN250 (PA, T, L) = .271 * PA ^ .824 * T ^ 1.079 * L: ' Eq. 5.11
510 DEF FNATTN500 (PA, T, L) = 1.77 * PA ^ .5 * T ^ 1.087 * L: ' Eq. 5.12
520 '
530 FOR I = 1 TO 6
540   FOR J = 1 TO 7
550     READ D(I, J): ' Reads in Absorption Coefficients
560   NEXT J
570 NEXT I
580 '
590 ' Absorption coefficients for various types of materials (Table 5.1)
600 '
610 DATA .01,.01,.01,.02,.02,.02,.03: ' Unpainted Poured Concrete
620 DATA .04,.04,.04,.05,.05,.05,.07: ' Bare Sheet Metal
630 DATA .02,.03,.22,.69,.91,.96,.99: ' 1" 3 lb/cu. ft. fiberglass insulation
640 DATA .18,.22,.82,1.21,1.1,1.02,1.05: ' 2" 3 lb/cu. ft. fiberglass insulation
650 DATA .48,.53,1.19,1.21,1.08,1.01,1.04: ' 3" 3 lb/cu. ft. fiberglass insulation
660 DATA .76,.84,1.24,1.24,1.08,1.0,0.97: ' 4" 3 lb/cu. ft. fiberglass insulation
670 '
680 CLS : PRINT "PLENUM SOUND ATTENUATION": PRINT
690 INPUT "ENTER INSIDE HEIGHT OF THE PLENUM (FT): ", PH
700 INPUT "ENTER INSIDE WIDTH OF THE PLENUM (FT): ", PW
710 INPUT "ENTER INSIDE LENGTH OF THE PLENUM (FT): ", PL
720 INPUT "ENTER HORIZONTAL DISTANCE BETWEEN CENTERS OF PLENUM INLET AND OUTLET (FT): ", PD
730 INPUT "ENTER VERTICAL DISTANCE BETWEEN CENTERS OF PLENUM INLET AND OUTLET (FT): ", PV
740 PRINT
750 PRINT "TYPE OF INLET OPENING:   (1. RECTANGULAR      2. CIRCULAR)"
760 INPUT "SELECTION: ", S1
770 IF S1 >= 1 AND S1 <= 2 THEN GOTO 780 ELSE GOTO 760
780 IF S1 = 2 THEN GOTO 830: ' S1 equals 2 when inlet is circular
790 INPUT "ENTER HEIGHT OF PLENUM INLET OPENING (IN): ", HI: HI = HI / 12
800 INPUT "ENTER WIDTH OF PLENUM INLET OPENING (IN): ", WI: WI = WI / 12
810 DI = 0
820 GOTO 850
830 INPUT "ENTER DIAMETER OF PLENUM INLET OPENING (IN): ", DI: DI = DI / 12
840 HI = 0: WI = 0
850 PRINT "TYPE OF OUTLET OPENING:   (1. RECTANGULAR      2. CIRCULAR)"
860 INPUT "SELECTION: ", S2
870 IF S2 >= 1 AND S2 <= 2 THEN GOTO 880 ELSE GOTO 860
880 IF S2 = 2 THEN GOTO 930: ' S2 equals 2 when outlet is circular
890 INPUT "ENTER HEIGHT OF PLENUM OUTLET OPENING (IN): ", HO: HO = HO / 12
900 INPUT "ENTER WIDTH OF PLENUM OUTLET OPENING (IN): ", WO: WO = WO / 12
910 OD = 0
920 GOTO 950
930 INPUT "ENTER DIAMETER OF PLENUM OUTLET OPENING (IN): ", OD: OD = OD / 12
940 HO = 0: WO = 0
950 PRINT
960 PRINT "NON-SOUND ABSORBING PLENUM SURFACE"
970 PRINT "1. CONCRETE"
980 PRINT "2. BARE SHEET METAL"
990 PRINT "3. NONE"
1000 INPUT "SELECTION: ", S3
1010 IF S3 >= 1 AND S3 <= 3 THEN GOTO 1020 ELSE GOTO 1000
1020 PRINT
1030 PRINT "SOUND ABSORBING PLENUM LINER"
1040 PRINT "1. 1 IN 3 LB/CU FT DENSITY FIBERGLASS INSULATION"
1050 PRINT "2. 2 IN 3 LB/CU FT DENSITY FIBERGLASS INSULATION"
1060 PRINT "3. 3 IN 3 LB/CU FT DENSITY FIBERGLASS INSULATION"
1070 PRINT "4. 4 IN 3 LB/CU FT DENSITY FIBERGLASS INSULATION"
1080 PRINT "5. NONE"
```

```
1090 INPUT "SELECTION: ", S4
1100 IF S4 >= 1 AND S4 <= 5 THEN GOTO 1110 ELSE GOTO 1090
1110 PRINT
1120 IF S4 = 1 THEN T = 1
1130 IF S4 = 2 THEN T = 2
1140 IF S4 = 3 THEN T = 3
1150 IF S4 = 4 THEN T = 4
1160 IF S3 = 3 THEN PC = 100: GOTO 1260: ' % Covered = 100%
1170 IF S4 = 5 THEN PC = 0: GOTO 1260: ' % Covered = 0%
1180 INPUT "ENTER % INSIDE SURFACE AREA COVERED BY SOUND ABSORBING PLENUM LINER: ", PC
1190 PRINT : PRINT "RADIATE SOUND INTO:   1. QUARTER SPACE      2. HALF SPACE"
1200 INPUT "SELECTION: ", S5
1210 IF S5 >= 1 AND S5 <= 2 THEN GOTO 1220 ELSE GOTO 1200
1220 PRINT
1230 PRINT "CONTINUE? (Y OR N)"
1240 Q$ = INKEY$: IF Q$ = "Y" OR Q$ = "N" OR Q$ = "y" OR Q$ = "n" THEN GOTO 1250 ELSE GOTO 1240
1250 IF Q$ = "N" OR Q$ = "n" THEN GOTO 680
1260 PRINT
1270 IF S4 = 5 THEN FOR I = 1 TO 7: TT(I) = D(S3, I): NEXT I
1280 IF S3 = 3 THEN FOR I = 1 TO 7: TT(I) = D(S4 + 2, I): NEXT I
1290 IF S3 <= 2 AND S4 <= 4 THEN FOR I = 1 TO 7: TT(I) = (PC * D(S4 + 2, I) + (100 - PC) * D(S
    I)) / 100: NEXT I
1300 RR = SQR(PD ^ 2 + PV ^ 2)
1310 TH = ATN(PV / PD): ' Calculation of theta (Eq. 4)
1320 IF S2 = 1 THEN SO = HO * WO: ' Outlet surface area for rectangular outlets, sq. ft.
1330 IF S2 = 2 THEN SO = PI * OD ^ 2 / 4: ' Outlet surface area for circular outlets, sq. ft.
1340 IF S1 = 1 THEN SI = HI * WI: ' Inlet surface area for rectangular inlets, sq. ft.
1350 IF S1 = 2 THEN SI = PI * DI ^ 2 / 4: ' Inlet surface area for circular inlets, sq. ft.
1360 IF S5 = 1 THEN Q = 4: ' Sound radiated into quarter space
1370 IF S5 = 2 THEN Q = 2: ' Sound radiated into half space
1380 TH = Q * COS(TH) / (4 * PI * RR ^ 2): ' Used in eq. 5.1 for parallel input and output
1390 AR = 2 * (PW * PH + PW * PL + PH * PL) - SI - SO: ' Total surface area, sq. ft.
1400 IF S1 = 1 AND HI > WI THEN COF = 1125 / (2 * HI): ' equation 5.5
1410 IF S1 = 1 AND WI >= HI THEN COF = 1125 / (2 * WI): ' equation 5.5
1420 IF S1 = 2 THEN COF = 659 / DI: ' equation 5.6
1430 PA = 2 * (PW + PH) / (PW * PH): ' P/A for use with eqs. 5.9-5.12
1440 SP = PW * PH
1450 M = SP / SI
1460 FOR I = 1 TO 7
1470 FR = 62.5 * 2 ^ (I - 1)
1480 IF FR < COF THEN GOTO 1530
1490 IF TT(I) > 1 THEN TT(I) = 1
1500 TT(I) = SO * (TH + (1 - TT(I)) / (AR * TT(I))): ' Eq. 5.1 respectively inside brackets
1510 T%(I) = -10 * LOG(TT(I)) / LOG(10): ' Equation 5.1
1520 GOTO 1600
1530 IF I = 1 THEN ATTN = FNATTN63(PA, T, PL): ' equation 5.9
1540 IF I = 2 THEN ATTN = FNATTN125(PA, T, PL): ' equation 5.10
1550 IF I = 3 THEN ATTN = FNATTN250(PA, T, PL): ' equation 5.11
1560 IF I = 4 THEN ATTN = FNATTN500(PA, T, PL): ' equation 5.12
1570 TLC = (FNCOSH(ATTN / 2) + .5 * (M + 1 / M) * FNSINH(ATTN / 2)) ^ 2 * (COS(2 * PI * FR * PL
    / 1125)) ^ 2
1580 TLS = (FNSINH(ATTN / 2) + .5 * (M + 1 / M) * FNCOSH(ATTN / 2)) ^ 2 * (SIN(2 * PI * FR * PL
    / 1125)) ^ 2
1590 T%(I) = 10 * LOG(TLC + TLS) / LOG(10): 'equation 5.7
1600 NEXT I
1610 '
1620 '   Output Statements
1630 '
1640 PRINT
1650 PRINT SPC(45); "1/1 OCT BAND CENTER FREQ"
1660 PRINT SPC(10); "DESCRIPTION"; SPC(21); " 63  125  250  500   1K   2K   4K"
1670 PRINT STRING$(41, CHR$(45)) + " ___ ___ ___ ___ ___ ___ ___"
1680 PRINT USING "\                                        \"; "PLENUM SOUND ATTENUATION, dB";
1690 PRINT SPC(6); USING " ###"; T%(1); T%(2); T%(3); T%(4); T%(5); T%(6); T%(7)
1700 END
```

SAMPLE OUTPUT FOR EXAMPLE 5.1

PLENUM SOUND ATTENUATION

ENTER INSIDE HEIGHT OF THE PLENUM (FT): 6
ENTER INSIDE WIDTH OF THE PLENUM (FT): 4
ENTER INSIDE LENGTH OF THE PLENUM (FT): 10
ENTER HORIZONTAL DISTANCE BETWEEN CENTERS OF PLENUM INLET AND OUTLET (FT): 10
ENTER VERTICAL DISTANCE BETWEEN CENTERS OF PLENUM INLET AND OUTLET (FT): 4

TYPE OF INLET OPENING: (1. RECTANGULAR   2. CIRCULAR)
SELECTION: 1
ENTER HEIGHT OF PLENUM INLET OPENING (IN): 24
ENTER WIDTH OF PLENUM INLET OPENING (IN): 36
TYPE OF OUTLET OPENING: (1. RECTANGULAR   2. CIRCULAR)
SELECTION: 1
ENTER HEIGHT OF PLENUM OUTLET OPENING (IN): 24
ENTER WIDTH OF PLENUM OUTLET OPENING (IN): 36

NON-SOUND ABSORBING PLENUM SURFACE
1. CONCRETE
2. BARE SHEET METAL
3. NONE
SELECTION: 2

SOUND ABSORBING PLENUM LINER
3. 1 IN 3 LB/CU FT DENSITY FIBERGLASS INSULATION
4. 2 IN 3 LB/CU FT DENSITY FIBERGLASS INSULATION
5. 3 IN 3 LB/CU FT DENSITY FIBERGLASS INSULATION
6. 4 IN 3 LB/CU FT DENSITY FIBERGLASS INSULATION
7. NONE
SELECTION: 1

ENTER % INSIDE SURFACE AREA COVERED BY SOUND ABSORBING PLENUM LINER: 100%

RADIATED SOUND INTO: 1. QUARTER SPACE   2. HALF SPACE
SELECTION: 1

CONTINUE? (Y OR N)

DESCRIPTION	63	125	250	500	1K	2K	4K
PLENUM SOUND ATTENUATION, dB	2	6	10	16	17	18	18

1/1 OCT BAND CENTER FREQ

UNLINED RECTANGULAR DUCTS

PROGRAM OUTLINE

Input system parameters:
   Duct dimensions
   Is duct externally insulated?
Calculate:
   Attenuation from Equations (5.13), (5.14) and (5.15)

COMPUTER PROGRAM – UNRDUC.BAS

```
10  '>UNLINED RECTANGULAR DUCT SOUND ATTENUATION COMPUTER PROGRAM<
20  '
30  DIM T%(8)
40  '
50  '     LIST OF VARIABLES
60  '
70  '     ATTN = total attenuation, dB
80  '     DH = larger duct dimension, in
90  '     DL = duct length, ft
100 '     DW = smaller duct dimension, in
110 '     FR = 1/1 octavwe band center frequency, Hz
120 '     PA = perimeter/area, 1/ft
130 '     S1 = selection for external insulation
140 '     T%() = 1/1 octave band attenuation output, dB
150 '
160 CLS : PRINT "SOUND ATTENUATION IN UNLINED RECTANGULAR SHEET METAL DUCTS": PRINT
170 INPUT "ENTER SMALL DUCT DIMENSION (IN): ", DW
180 INPUT "ENTER LARGER DUCT DIMENSION (IN): ", DH
190 INPUT "ENTER DUCT LENGTH (FT): ", DL
200 PRINT "IS DUCT EXTERNALLY INSULATED?  (1. YES    2. NO)"
210 INPUT "SELECTION: ", S1
220 IF S1 >= 1 AND S1 <= 2 THEN GOTO 230 ELSE GOTO 210
230 PRINT
240 PRINT "CONTINUE? (Y OR N)"
250 Q$ = INKEY$: IF Q$ = "Y" OR Q$ = "N" OR Q$ = "y" OR Q$ = "n" THEN GOTO 260 ELSE GOTO 250
260 IF Q$ = "N" OR Q$ = "n" THEN GOTO 160
270 '
280 PA = 24 * (DW + DH) / (DW * DH): ' Perimenter/Area (1/ft)
290 FOR I = 1 TO 8
300 FR = 62.5 * 2 ^ (I - 1)
310 IF I <= 3 AND PA >= 3 THEN ATTN = 17 * PA ^ (-.25) * FR ^ (-.85) * DL: ' equation (5.13)
320 IF I <= 3 AND PA < 3 THEN ATTN = 1.64 * PA ^ .73 * FR ^ (-.58) * DL: ' equation (5.14)
330 IF I > 3 THEN ATTN = .02 * PA ^ .8 * DL: ' equation (5.15)
340 IF I <= 3 AND S1 = 1 THEN ATTN = 2 * ATTN: ' externally lined duct
350 T%(I) = ATTN: ' Total Attenuation
360 NEXT I
370 '
380 ' Output statements
390 '
400 PRINT
410 PRINT SPC(45); "1/1 OCT BAND CENTER FREQ"
420 PRINT SPC(10); "DESCRIPTION"; SPC(21); " 63  125  250  500  1K   2K   4K   8K"
430 PRINT STRING$(41, CHR$(45)) + " ___  ___  ___  ___  ___  ___  ___  ___"
440 PRINT USING "\                                        \"; "UNLINED RECT. DUCT SOUND ATTENUATION, dB";
450 PRINT USING "### "; T%(1); T%(2); T%(3); T%(4); T%(5); T%(6); T%(7); T%(8)
```

SAMPLE OUTPUT FOR EXAMPLE 5.2

SOUND ATTENUATION IN UNLINED RECTANGULAR SHEET METAL DUCTS

ENTER SMALL DUCT DIMENSION (IN): 12
ENTER LARGER DUCT DIMENSTION (IN): 18
ENTER DUCT LENGTH (FT): 20
IS DUCT EXTERNALLY INSULATED? (1. YES   2. NO)
SELECTION: 2

CONTINUE? (Y OR N)

DESCRIPTION	1/1 OCT BAND CENTER FREQ							
	63	125	250	500	1K	2K	4K	8K
UNLINED RECT. DUCT SOUND ATTEN. dB	7	4	2	1	1	1	1	1

ACOUSTICALLY LINED RECTANGULAR DUCTS

PROGRAM OUTLINE

Input data from Table 5.8
Input system parameters:
    Duct dimensions
    Thickness of fiberglass lining
Calculate:
    P/A, perimeter divided by the area
    Insertion loss from equation 5.16
    natural attenuation from equations 5.13, 5.14 and 5.15
    Total attenuation from equation 5.17

COMPUTER PROGRAM – LINRDUCT.BAS

```
10  '>INSERTION LOSS OF ACOUSTICALLY LINED RECTANGULAR DUCTS COMPUTER PROGRAM<
20  '
30  DIM C1(8), C2(8), C3(8), T%(8)
40  '
50  '      LIST OF VARIABLES
60  '
70  '      ATTN = natural attenuation, dB (eqs. 5.13-5.15)
80  '      C1() = the coefficient A for use in equation 5.16
90  '      C2() = the coefficient B for use in equation 5.16
100 '      C3() = the coefficient C for use in equation 5.16
110 '      DH = smaller inside duct dimension, in
120 '      DL = duct length, ft
130 '      DW = smaller inside duct dimension, in
140 '      FR = 1/1 octave band center frequency
150 '      IL = insertion loss, dB (eq. 5.16)
160 '      PA = perimeter/area, 1/ft
170 '      S1 = selection for thickness of lining
180 '      T = thickness of lining, in
190 '      T%() = total sound attenuation, dB
200 FOR I = 1 TO 8: READ C1(I): NEXT I: ' Reads in c1, c2, and c3
210 FOR I = 1 TO 8: READ C2(I): NEXT I: ' for method based on
220 FOR I = 1 TO 8: READ C3(I): NEXT I: ' Product data.
230 '
240 CLS : PRINT "SOUND ATTENUATION OF ACOUSTICALLY LINED RECTANGULAR DUCTS": PRINT
250 PRINT "THIS ANALYSIS IS FOR RECTANGULAR DUCTS WITH FIBERGLASS LININGS THAT HAVE A"
260 PRINT "DENSITY OF 1.5-3.0 LB/CU FT.": PRINT
270 '
280 ' Input Statements
290 '
300 INPUT "ENTER SMALLER INSIDE DUCT DIMENSION (IN): ", DW
310 INPUT "ENTER LARGER INSIDE DUCT DIMENSION (IN): ", DH
320 INPUT "ENTER DUCT LENGTH (FT): ", DL
330 PRINT "1. 1 IN DUCT LINING"; SPC(11); "2. 2 IN DUCT LINING"
340 INPUT "SELECTION: ", S1
350 IF S1 >= 1 AND S1 <= 2 THEN GOTO 360 ELSE GOTO 340
360 PRINT
370 PRINT "CONTINUE? (Y OR N)"
380 Q$ = INKEY$: IF Q$ = "Y" OR Q$ = "N" OR Q$ = "y" OR Q$ = "n" THEN GOTO 390 ELSE GOTO 380
390 IF Q$ = "N" OR Q$ = "n" THEN GOTO 240
400 '
410 ' Equations for attenuation based on product data come from regression analysis using
420 ' Data from several manufacturers.
430 '
440 ' PA = Perimeter/Area in 1/ft
450 '
460 PA = 24 * (DW + DH) / (DW * DH)
470 '
480 T = S1: ' T = thickness in inches
490 '
500 FOR I = 1 TO 8
510 '
```

```
520     ' Calculates 1/1 octave band center frequency
530     '
540     FR = 62.5 * 2 ^ (I - 1)
550     '
560     ' Regression equations for attenuation based on product data
570     ' C1, C2, and C3 are constants which are read in using data statements
580     '
590     IL = C1(I) * PA ^ C2(I) * T ^ C3(I) * DL: ' equation 5.16
600     '
610     ' Calculate the natural attenuation of rectangular ducts
620     '
630     IF I <= 3 AND PA >= 3 THEN ATTN = 17 * PA ^ (-.25) * FR ^ (-.85) * DL: ' equation 5.13
640     IF I <= 3 AND PA < 3 THEN ATTN = 1.64 * PA ^ .73 * FR ^ (-.58) * DL: ' equation 5.14
650     IF I > 3 THEN ATTN = .02 * PA ^ .8 * DL: ' equation 5.15
660     '
670     ' Calculate the total attenuation of rectangular lined ducts
680     '
690     T%(I) = IL + ATTN: ' equation 5.17
700     '
710     ' To prevent the prediction of excessive values, the insertion loss in
720     ' any straight, lined section should be limited to 40 dB at any frequency
730     '
740     IF T%(I) > 40 THEN T%(I) = 40
750 NEXT I
760     '
770     ' Output statements
780     '
790     V$ = STR$(V)
800     '
810 PRINT
820 PRINT SPC(45); "1/1 OCT BAND CENTER FREQ"
830 PRINT SPC(10); "DESCRIPTION"; SPC(21); " 63  125  250  500   1K   2K   4K   8K"
840 PRINT STRING$(41, CHR$(45)) + " ___ ___ ___ ___ ___ ___ ___ ___"
850 PRINT USING "\                                        \"; "LINED DUCT SOUND ATTENUATION, dB";
860 PRINT USING " ###"; T%(1); T%(2); T%(3); T%(4); T%(5); T%(6); T%(7); T%(8)
870 DATA 0.0133,0.0574,0.271,1.0147,1.770,1.392,1.518,1.581
880 DATA 1.959,1.410,0.824,0.500,0.695,0.802,0.451,0.219
890 DATA 0.917,0.941,1.079,1.087,0.0,0.0,0.0,0.0
```

SAMPLE OUTPUT FOR EXAMPLE 5.3

SOUND ATTENUATION OF ACOUSTICALLY LINED RECTANGULAR DUCTS

THIS ANALYSIS IS FOR RECTANGULAR DUCTS WITH FIBERGLASS LININGS THAT HAVE A DENSITY OF 1.5–3.0 LB/CU FT.

```
ENTER SMALLER INSIDE DUCT DIMENSION (IN): 24
ENTER LARGER INSIDE DUCT DIMENSION (IN): 36
ENTER DUCT LENGTH (FT): 10
1. 1 IN DUCT LINING            2. 2 IN DUCT LINING
SELECTION: 1
```

CONTINUE? (Y OR N)

| | 1/1 OCT BAND CENTER FREQ ||||||||
DESCRIPTION	63	125	250	500	1K	2K	4K	8K
LINED DUCT SOUND ATTENUATION, dB	3	3	5	14	26	21	19	18

COMPUTER PROGRAMS FOR CHAPTER 5

UNLINED CIRCULAR DUCTS

PROGRAM OUTLINE

Input system parameters:
    Duct diameter, in
    Duct length, ft
Calculate:
    Attenuation, dB/ft (Table 5.5)
    Total Attenuation, dB

COMPUTER PROGRAM – UNCDUC.BAS

```
10  'Duct Element Sound Attenuation:  Unlined Circular Ducts (4/20/88)
20  '
30  DIM T%(8)
40  PI = 3.141592654#
50  '
60  '     LIST OF VARIABLES
70  '
80  '     DD = duct diameter, in
90  '     DL = duct length, ft
100 '     TT() = attenuation, dB/ft
110 '     T%() = total attenuation, dB
120 '
130 CLS : PRINT "SOUND ATTENUATION IN UNLINED CIRCULAR DUCTS": PRINT " "
140 '
150 'Input Statements for Duct Dimensions and Volume Flow
160 '
170 INPUT "ENTER DUCT DIAMETER (IN):", DD
180 INPUT "ENTER DUCT LENGTH (FT):", DL
190 PRINT : PRINT "CONTINUE? (Y OR N)"
200 Q$ = INKEY$: IF Q$ = "Y" OR Q$ = "N" OR Q$ = "y" OR Q$ = "n" THEN GOTO 210 ELSE GOTO 200
210 IF Q$ = "N" OR Q$ = "n" THEN GOTO 130
220 '
230 'Approximate sound power attenuation (dB/ft), which is dependent upon diameter
240 '
250 IF DD <= 7 THEN TT(1) = .03: TT(2) = .03: TT(3) = .05: TT(4) = .05: TT(5) = .1: TT(6) = .1: TT(7) = .1: TT(8) = .1
260 IF DD > 7 AND DD <= 15 THEN TT(1) = .03: TT(2) = .03: TT(3) = .03: TT(4) = .05: TT(5) = .07: TT(6) = .07: TT(7) = .07: TT(8) = .07
270 IF DD > 15 AND DD <= 30 THEN TT(1) = .02: TT(2) = .02: TT(3) = .02: TT(4) = .05: TT(5) = .05: TT(6) = .05: TT(7) = .05: TT(8) = .05
280 IF DD > 30 THEN TT(1) = .01: TT(2) = .01: TT(3) = .01: TT(4) = .02: TT(5) = .02: TT(6) = .02: TT(7) = .02: TT(8) = .02
290 '
300 'Loop for determining sound power attenuation for all 8 1/1 Octave bands
310 '
320 FOR I = 1 TO 8
330 T%(I) = TT(I) * DL: ' Total Attenuation
340 NEXT I
350 '
360 '     Output Statements
370 '
380 PRINT
390 PRINT SPC(45); "1/1 OCT BAND CENTER FREQ"
400 PRINT SPC(10); "DESCRIPTION"; SPC(21); "  63  125  250  500   1K   2K   4K   8K"
410 PRINT STRING$(41, CHR$(45)) + " ---- ---- ---- ---- ---- ---- ---- ----"
420 PRINT USING "\                                        \"; "UNLINED CIR. DUCT SOUND ATTENUATION, dB";
430 PRINT USING " ###"; T%(1); T%(2); T%(3); T%(4); T%(5); T%(6); T%(7); T%(8)
```

SAMPLE OUTPUT FOR EXAMPLE 5.4

SOUND ATTENUATION IN UNLINED CIRCULAR DUCTS

ENTER DUCT DIAMETER (IN): 12
ENTER DUCT LENGTH (FT): 20

CONTINUE? (Y OR N)

DESCRIPTION	63	125	250	500	1K	2K	4K	8K
UNLINED CIR. DUCT SOUND ATT., dB	1	1	1	1	1	1	1	1

(1/1 OCT BAND CENTER FREQ)

ACOUSTICALLY LINED CIRCULAR DUCTS

PROGRAM OUTLINE

Input system Parameters
    Inside duct diameter, in
    Duct lining thickness, in
    Duct length, ft
Calculate:
    Total insertion loss, IL (Eq. 5.18)

COMPUTER PROGRAM – LINCDUCT.BAS

```
10  '>LINED CIRCULAR DUCT SOUND ATTENUATION SUBROUTINE<
20  '
30  PI = 3.14159264#
40  DIM T%(8)
50  '
60  '    LIST OF VARIABLES
70  '
80  '    A() = coefficient for use with equation 1 respectively
90  '    B() = coefficient for use with equation 1 respectively
100 '    C() = coefficient for use with equation 1 respectively
110 '    D   = inside duct diameter, in
120 '    DL  = duct length, th
130 '    D() = coefficient for use with equation 1 respectively
140 '    E() = coefficient for use with equation 1 respectively
150 '    F() = coefficient for use with equation 1 respectively
160 '    TH  = thickness of acoustical lining, in
170 '    T%() = total sound attenuation, dB
180 '
190 CLS : PRINT "SOUND ATTENUATION IN CIRCULAR DUCTS BASED ON MEASURED DATA": PRINT
200 PRINT "ACOUSTICAL LINING OF DUCT - 0.75 LB/CU FT DENSITY FIBERGLASS"
210 PRINT "DUCT - DUAL WALL SPIRAL SHEET METAL DUCT"
220 PRINT "INNER LINER - PERFERATED SHEET METAL WITH 25% OPEN AREA": PRINT
230 PRINT "VALID FOR INSIDE DUCT DIAMETERS OF 6 TO 60 INCHES"
240 PRINT
250 INPUT "ENTER INSIDE DUCT DIAMETER (IN): ", D: IF D > 60 OR D < 6 THEN GOTO 250
260 PRINT
270 PRINT "VALID FOR LINING THICKNESSES OF 1 TO 3 INCHES"
280 PRINT
290 INPUT "ENTER DUCT LINING THICKNESS (IN): ", TH: IF TH > 3 OR TH < 1 THEN GOTO 290
300 PRINT
310 INPUT "ENTER DUCT LENGTH (FT): ", DL
320 PRINT
330 PRINT "CONTINUE? (Y OR N)"
340 Q$ = INKEY$: IF Q$ = "Y" OR Q$ = "N" OR Q$ = "y" OR Q$ = "n" THEN GOTO 350 ELSE GOTO 340
350 IF Q$ = "N" OR Q$ = "n" THEN GOTO 190
360 '
370 GOSUB 570
380 '
390 FOR I = 1 TO 8
400 ' Equation 1
410 T%(I) = DL * (A(I) + B(I) * TH + C(I) * TH ^ 2 + D(I) * D + E(I) * D ^ 2 + F(I) * D ^ 3)
420 IF T%(I) > 40 THEN T%(I) = 40
430 NEXT I
440 '
450 ' Output statements
460 '
470 PRINT
480 PRINT SPC(45); "1/1 OCT BAND CENTER FREQ"
490 PRINT SPC(10); "DESCRIPTION"; SPC(21); " 63 125 250 500  1K  2K  4K  8K"
500 PRINT STRING$(41, CHR$(45)) + " ___ ___ ___ ___ ___ ___ ___ ___"
510 PRINT USING "\                                        \"; "CIRCULAR LINED DUCT SOUND ATTENUATION, dB";
520 PRINT USING " ###"; T%(1); T%(2); T%(3); T%(4); T%(5); T%(6); T%(7); T%(8);
530 END
540 '
550 ' Subroutine for reading in coefficients for regression equations
```

```
560 '
570 FOR I = 1 TO 8
580    READ A(I), B(I), C(I), D(I), E(I), F(I)
590 NEXT I
600 '
610 '
620 ' Data for regression equations
630 '
640 ' Data low frequency insertion loss
650 DATA 0.2825,  0.3447, -5.251E-2, -0.03837,9.131E-4, -8.294E-6: ' 63 Hz
660 DATA 0.5237,0.2234,-4.936E-3,-0.02724,3.377E-4, -2.469E-6: ' 125 Hz
670 DATA 0.3652,0.7900,-0.1157,-1.834E-2,-1.211E-4, 2.681E-6: ' 250 Hz
680 DATA 0.1333,1.845, -0.3735, -1.293E-2,8.624E-5, -4.986E-6: ' 500 Hz
690 ' Data for high frequency insertion loss
700 DATA 1.933, 0.0, 0.0, 6.135E-2,  -3.891E-3, 3.934E-5: ' 1K Hz
710 DATA 2.730, 0.0, 0.0, -7.341E-2, 4.428E-4,  1.006E-6: ' 2K Hz
720 DATA 2.800, 0.0, 0.0, -0.1467,   3.404E-3,  -2.851E-5: ' 4K Hz
730 DATA 1.545, 0.0, 0.0, -5.452E-2, 1.290E-3,  -1.318E-5: ' 8K Hz
740 '
750 RETURN
```

SAMPLE OUTPUT FOR EXAMPLE 5.5

SOUND ATTENUATION IN CIRCULAR DUCTS BASED ON MEASURED DATA

ACOUSTICAL LINING OF DUCT — 0.75 LB/CU FT DENSITY FIBERGLASS
DUCT — DUAL WALL SPIRAL SHEET METAL DUCT
INNER LINER — PERFERATED SHEET METAL WITH 25% OPEN AREA

VALID FOR INSIDE DUCT DIAMETERS OF 6 TO 60 INCHES

ENTER INSIDE DUCT DIAMETER (in): 24

VALID FOR LINING THICKNESSES OF 1 TO 3 INCHES

ENTER DUCT LINING THICKNESS (in): 1

ENTER DUCT LENGTH (ft): 10

CONTINUE? (Y OR N)

DESCRIPTION	63	125	250	500	1K	2K	4K	8K
CIRCULAR LINED DUCT SOUND ATTENUATION, dB	1	2	6	13	17	12	8	8

1/1 OCT BAND CENTER FREQ

ELBOWS

PROGRAM OUTLINE

Input system parameters:
- Type of elbow
- Insulation
- With or without turning vanes
- Duct width in plane of bend or duct diameter, in

Calculate:
- Determine values from Tables 5.13 through 5.15
- BW
- Insertion loss values

COMPUTER PROGRAM – ELBOWS.BAS

```
10 REM >LINED RECTANGULAR ELBOWS SUBROUTINE<
20 DIM T%(8)
30 '
40 '     LIST OF VARIABLES
50 '
60 '     A1-A6 = data corresponding to Tables 5.13 through 5.15
70 '     BW = frequency x width/ 1000 (Eq. 5.19)
80 '     FR = 1/1 octave band center frequency, Hz
90 '     H = duct width in plane of bend or duct diameter, in
100 '    S1 = selection for geometry of elbow
110 '    S2 = selection for lined or unlined elbows
120 '    S3 = selection for turning vanes
130 '
140 CLS : PRINT "SOUND ATTENUATION OF ELBOWS": PRINT
150 '
160 ' Select type of elbow
170 '
180 PRINT "1. SQUARE ELBOW              2. ROUND ELBOW"
190 INPUT "SELECTION: ", S1
200 IF S1 >= 1 AND S1 <= 2 THEN GOTO 210 ELSE GOTO 190
210 IF S1 = 1 THEN PRINT "1. UNLINED                 2. LINED"
220 IF S1 = 1 THEN INPUT "SELECTION: ", S2
230 IF S1 = 1 THEN GOTO 240 ELSE GOTO 250
240 IF S2 >= 1 AND S2 <= 2 THEN GOTO 250 ELSE GOTO 220
250 IF S1 = 1 THEN PRINT "1. W/O TURNING VANES       2. WITH TURNING VANES"
260 IF S1 = 1 THEN INPUT "SELECTION: ", S3
270 IF S1 = 1 THEN GOTO 280 ELSE GOTO 290
280 IF S3 >= 1 AND S3 <= 2 THEN GOTO 290 ELSE GOTO 260
290 '
300 ' Input demension in plane of bend or duct diameter
310 '
320 PRINT
330 INPUT "ENTER DUCT WIDTH IN PLANE OF BEND OR DUCT DIAMETER (IN): ", H
340 PRINT
350 PRINT "CONTINUE? (Y OR N)"
360 Q$ = INKEY$: IF Q$ = "Y" OR Q$ = "N" OR Q$ = "y" OR Q$ = "n" THEN GOTO 370 ELSE GOTO 360
370 IF Q$ = "N" OR Q$ = "n" THEN GOTO 140
380 '
390 ' Set values of insertion loss values based on input selections
400 '
410 ' Data from Table 5.13
420 '
430 IF S1 = 1 AND S2 = 1 AND S3 = 1 THEN A1 = 0: A2 = 1: A3 = 5: A4 = 8: A5 = 4: A6 = 3
440 IF S1 = 1 AND S2 = 2 AND S3 = 1 THEN A1 = 0: A2 = 1: A3 = 6: A4 = 11: A5 = 10: A6 = 10
450 '
460 ' Data from Table 5.14
470 '
480 IF S1 = 1 AND S2 = 1 AND S3 = 2 THEN A1 = 0: A2 = 1: A3 = 4: A4 = 6: A5 = 4: A6 = 4
490 IF S1 = 1 AND S2 = 2 AND S3 = 2 THEN A1 = 0: A2 = 1: A3 = 4: A4 = 7: A5 = 7: A6 = 7
500 '
510 ' Data from Table 5.15
```

```
520 '
530 IF S1 = 2 THEN A1 = 0: A2 = 1: A3 = 2: A4 = 3: A5 = 3: A6 = 3
540 FOR I = 1 TO 8
550 FR = 62.5 * 2 ^ (I - 1): ' Calculate value for frequency
560 BW = FR * H / 1000: ' equation (1)
570 '
580 ' Calculate values of insertion loss
590 '
600 IF BW < 1.9 THEN T%(I) = A1
610 IF BW >= 1.9 AND BW < 3.8 THEN T%(I) = A2
620 IF BW >= 3.8 AND BW < 7.5 THEN T%(I) = A3
630 IF BW >= 7.5 AND BW < 15 THEN T%(I) = A4
640 IF BW >= 15 AND BW < 30 THEN T%(I) = A5
650 IF BW >= 30 THEN T%(I) = A6
660 NEXT I
670 '
680 ' Output statements
690 '
700 ' Select lable
710 '
720 IF S1 = 1 AND S2 = 1 AND S3 = 1 THEN T1$ = "UNL SQ ELBOW W/O TV SOUND ATTEN, dB"
730 IF S1 = 1 AND S2 = 2 AND S3 = 1 THEN T1$ = "LIN SQ ELBOW W/O TV SOUND ATTEN, dB"
740 IF S1 = 1 AND S2 = 1 AND S3 = 2 THEN T1$ = "UNL SQ ELBOW WITH TV SOUND ATTEN, dB"
750 IF S1 = 1 AND S2 = 2 AND S3 = 2 THEN T1$ = "LIN SQ ELBOW WITH TV SOUND ATTEN, dB"
760 IF S1 = 2 THEN T1$ = "ROUND ELBOW SOUND ATTENUATION, dB"
770 PRINT
780 PRINT SPC(45); "1/1 OCT BAND CENTER FREQ"
790 PRINT SPC(10); "DESCRIPTION"; SPC(21); " 63  125  250  500   1K   2K   4K   8K"
800 PRINT STRING$(41, CHR$(45)) + "  ___  ___  ___  ___  ___  ___  ___  ___"
810 PRINT USING "\                                        \"; T1$;
820 PRINT USING "### "; T%(1); T%(2); T%(3); T%(4); T%(5); T%(6); T%(7); T%(8)
830 END
```

SAMPLE OUTPUT FOR EXAMPLE 5.6

SOUND ATTENUATION OF ELBOWS

1. SQUARE ELBOW   2. ROUND ELBOW
SELECTION: 1
1. UNLINED   2. LINED
SELECTION: 2
1. W/O TURNING VANES   2. WITH TURNING VANES
SELECTION: 1

ENTER DUCT WIDTH IN PLANE OF BEND OR DUCT DIAMETER (IN): 24

CONTINUE? (Y OR N)

DESCRIPTION	63	125	250	500	1K	2K	4K	8K
LIN SQ ELBOW W/O TV SOUND ATTEN, dB	0	1	6	11	10	10	10	10

SAMPLE OUTPUT FOR EXAMPLE 5.7

SOUND ATTENUATION OF ELBOWS

1. SQUARE ELBOW   2. ROUND ELBOW
SELECTION: 2

ENTER DUCT WIDTH IN PLANE OF BEND OR DUCT DIAMETER (IN): 12

CONTINUE? (Y OR N)

DESCRIPTION	63	125	250	500	1K	2K	4K	8K
ROUND ELBOW SOUND ATTENUATION, dB	0	0	1	2	3	3	3	3

ACOUSTICALLY LINED CIRCULAR RADIUSED ELBOWS

PROGRAM OUTLINE

Input System Parameters:
   Duct inside diameter, d (inches)
   radius of elbow (to centerline of duct), r (inches)
      Default is equation 5.21
   Lining thickness, t (inches)
Calculate:
   1/1 octave band center frequency
   f·d
   $IL \cdot \left[\frac{d}{r}\right]^2$ (dB)
   Insertion loss, IL (dB)

COMPUTER PROGRAM – LINCELB.BAS

```
10  '>LINED CIRCULAR ELBOW SOUND ATTENUATION COMPUTER PROGRAM<
20  '
30  DIM T%(8)
40  '
50  '     LIST OF VARIABLES
60  '
70  '     D = inside duct diameter, in
80  '     FD = frequency x diameter
90  '     FZ = 1/1 octave band center frequency
100 '     R = radius of elbow (to centerline of duct)
110 '     TH = thickness of acoustical lining, in
120 '     TT = insertion loss x (R/D)^2
130 '     T%() = total sound attenuation, dB
140 '
150 CLS : PRINT "SOUND ATTENUATION IN CIRCULAR ELBOWS BASED ON MEASURED DATA": PRINT
160 PRINT "ACOUSTICAL LINING OF ELBOW – 0.75 LB/CU FT DENSITY FIBERGLASS"
170 PRINT "DUCT – DUAL WALL SPIRAL SHEET METAL DUCT"
180 PRINT "INNER LINER – PERFORATED SHEET METAL WITH 25% OPEN AREA": PRINT
190 PRINT "VALID FOR INSIDE DIAMETERS OF 6 TO 60 INCHES"
200 PRINT
210 INPUT "ENTER INSIDE DUCT DIAMETER (in): ", D: IF D > 60 OR D < 6 THEN GOTO 210
220 PRINT
230 PRINT "ENTER RADIUS OF ELBOW (TO CENTERLINE OF DUCT)"
240 INPUT "(DEFAULT = 1.5xdiameter + 3xthickness): ", R
250 PRINT
260 PRINT "VALID FOR LINING THICKNESSES OF 1 TO 3 INCHES"
270 PRINT
280 INPUT "ENTER DUCT LINING THICKNESS (in): ", TH: IF TH > 3 OR TH < 1 THEN GOTO 280
290 PRINT
300 PRINT "CONTINUE? (Y OR N)"
310 Q$ = INKEY$: IF Q$ = "Y" OR Q$ = "N" OR Q$ = "y" OR Q$ = "n" THEN GOTO 320 ELSE GOTO 310
320 IF Q$ = "N" OR Q$ = "n" THEN GOTO 150
330 '
340 '
350  IF R = 0! THEN R = 1.5 * D + 3 * TH: ' Equation 5.21
360 '
370 FOR I = 1 TO 8
380    FZ = 62.5 * 2 ^ (I – 1): ' 1/1 octave band center frequency, Hz
390    FD = LOG(D * FZ / 1000) / LOG(10): ' frequency x diameter
400 '  Equation 5.19
410    IF D <= 18 THEN TT = .485 + 2.094 * FD + 3.172 * FD ^ 2 – 1.578 * FD ^ 4 + .085 * FD ^ 7
420 '  Equation 5.20
430    IF D > 18 THEN TT = –1.493 + .538 * TH + 1.406 * FD + 2.779 * FD ^ 2 – .662 * FD ^ 4 + .016 * FD ^ 7
440    T%(I) = TT / ((D / R) ^ 2): ' Insertion loss in dB
450    IF T%(I) > 30 THEN T%(I) = 30: ' Limits insertion loss to 30 dB
460 NEXT I
470 '
480 ' Output statements
```

```
490 '
500 PRINT
510 PRINT SPC(45); "1/1 OCT BAND CENTER FREQ"
520 PRINT SPC(10); "DESCRIPTION"; SPC(21); "  63 125 250 500  1K  2K  4K  8K"
530 PRINT STRING$(41, CHR$(45)) + "  ___ ___ ___ ___ ___ ___ ___ ___"
540 PRINT USING "\                                            \"; "SOUND ATTEN FOR LINED CIRC ELBOWS, dB";
550 PRINT USING "  ###"; T%(1); T%(2); T%(3); T%(4); T%(5); T%(6); T%(7); T%(8);
560 END
```

SAMPLE OUTPUT FOR EXAMPLE 5.8

SOUND ATTENUATION IN CIRCULAR ELBOWS BASED ON MEASURED DATA

ACOUSTICAL LINING OF ELBOW — 0.75 LB/CU FT DENSITY FIBERGLASS
DUCT — DUAL WALL SPIRAL SHEET METAL DUCT
INNER LINER — PERFERATED SHEET METAL WITH 25% OPEN AREA

VALID FOR INSIDE DIAMETERS OF 6 TO 60 INCHES

ENTER INSIDE DUCT DIAMETER (in): 24

ENTER RADIUS OF ELBOW (TO CENTERLINE OF DUCT)
(DEFAULT = 1.5 x diameter + 3 x thickness):

VALID FOR LINING THICKNESSES OF 1 TO 3 INCHES

ENTER DUCT LINING THICKNESS (in): 2

CONTINUE? (Y OR N)

DESCRIPTION	63	125	250	500	1K	2K	4K	8K
SOUND ATTEN FOR LINED CIRC ELBOWS, dB	0	3	6	11	14	16	15	14

(header above columns: 1/1 OCT BAND CENTER FREQ)

DUCT SILENCERS

PROGRAM OUTLINE

Select type of duct silencer
Read in data
Input system parameters:
    1/1 octave band sound power levels before silencer
    silencer cross section
    number of silencers to make up total face area
    Duct volume flow rate, cfm
    Direction of sound propagation
Calculate:
    Silencer face velocity, fpm
    Static pressure drop across silencer, in H_2O
    Duct insertion loss, dB
    Airflow generated noise, dB
    Sound power level after silencer, dB

COMPUTER PROGRAM – DUCTSIL.BAS

```
10  ' >DUCT SILENCER COMPUTER PROGRAM<
20  DIM T$(60), T%(8), D(10, 42, 10), L$(8, 25), FV(8, 31), SP(8, 11), PWL(8)
30  DIM DA(8), DB(8), DC(8), DD(8), DIL(8, 8), DILF(8), AFGN(8, 8), AFGNF(8)
40  CLS : PRINT "SOUND ATTENUATION AND REGENERATED NOISE OF DUCT SILENCERS": PRINT
50  '
60  '   List of Variables
70  '
80  '   ADJ =       factor used for linear interpolation
90  '   AFGN()  =   Airflow generated noise, dB
100 '   AFGNF() =   Airflow generated noise after interpolation
110 '   AR =        face area
120 '   SP() =      contains coefficients used for static pressure drop
130 '   FV() =      contains flow velocities corresponding to data
140 '   CUP =       system coefficient for upstream system component
150 '   CDOWN =     system coefficient for downstream system component
160 '   C5 =        system coefficient for both system components
170 '   D() =       Data for duct insertion loss and airflow generated noise
180 '   DA() =      number of flow velocities for DIL and AFGN values
190 '   DB() =      number of labels
200 '   DC() =      number of sets of data for DIL and AFGN
210 '   DD() =      number of DIL values in each flow direction
220 '   DEQ =       equivalent duct diameter (ft)
230 '   DIA =       Silencer diameter, in
240 '   DIL() =     Duct insertion loss, dB
250 '   DILF() =    Duct insertion loss after interpolation
260 '   DOWNDIST =  distance between silencer and downstream system component (ft)
270 '   FAAF =      Face area adjustment factor, dB
280 '   FV =        Duct volume flow rate, cfm
290 '   FZ =        1/1 Octave band center frequency
300 '   L$() =      Labels used for menus given duct sizes
310 '   LW2 =       see equation 4.30
320 '   NDEQ =      no. of equiv. duct diam. between silencer and system component
330 '   NS =        number of silencers
340 '   N1 =        used for DIL values
350 '   N2 =        used for AFGN values
360 '   PD =        Static pressure drop, inches H20
370 '   PDC =       used for pressure drop in circular silencers to simplify equation
380 '   PWL() =     Lw1, original sound power level
390 '   S1 =        selection for rectangular or circular ducts
400 '   S3 =        selection for size of silencer
410 '   S4 =        selection for upstream system components
420 '   S5 =        selection for downstream system components
430 '   S6 =        selection for with or against air flow
440 '   T%() =      Lw4, Sound power after attenuation, eq. 4.32
```

COMPUTER PROGRAMS FOR CHAPTER 5

```
450 '    UPDIST =    distance between silencer and upstream system component (ft)
460 '    VA =        flow velocity, fpm
470 '    VL =        flow velocity, fpm
480 '
490 PRINT "INPUT SOUND POWER LEVELS IN DUCT BEFORE SILENCER"
500 FOR I = 1 TO 8
510   FZ = 62.5 * 2 ^ (I - 1)
520   PRINT "SOUND POWER LEVEL FOR THE"; FZ; " HZ OCTAVE BAND "; : INPUT PWL(I)
530 NEXT I
540 PRINT : FLAG = 1
550 PRINT "CONTINUE? (Y OR N)"
560 Q$ = INKEY$: IF Q$ = "Y" OR Q$ = "N" OR Q$ = "y" OR Q$ = "n" THEN GOTO 570 ELSE GOTO 560
570 IF Q$ = "N" OR Q$ = "n" THEN GOTO 40
580 CLS
590 PRINT "RECTANGULAR DUCT SILENCERS"
600 PRINT "1. STANDARD PRESSURE DROP"
610 PRINT "2. LOW PRESSURE DROP"
620 PRINT
630 PRINT "CIRCULAR DUCT SILENCERS, INSULATED CONE AND OUTER SURFACE"
640 PRINT "3. STANDARD PRESSURE DROP"
650 PRINT "4. LOW PRESSURE DROP"
660 INPUT "SELECTION: ", S1
670 IF S1 >= 1 AND S1 <= 4 THEN GOTO 680 ELSE GOTO 660
680 PRINT
690 IF FLAG <> 1 THEN GOTO 990
700 FOR I = 1 TO 4
710   READ DA(I), DB(I), DC(I), DD(I)
720 NEXT I
730 FOR I = 1 TO 4
740   FOR J = 1 TO DC(I)
750     FOR K = 1 TO 8
760       READ D(I, J, K)
770     NEXT K
780   NEXT J
790 NEXT I
800 '
810 FOR I = 1 TO 4
820   FOR J = 1 TO DB(I)
830     READ L$(I, J)
840   NEXT J
850 NEXT I
860 '
870 FOR I = 1 TO 4
880   FOR J = 1 TO 4
890     READ SP(I, J)
900   NEXT J
910 NEXT I
920 '
930 FOR I = 1 TO 4
940   FOR J = 1 TO DA(I)
950     READ FV(I, J)
960   NEXT J
970 NEXT I
980 '
990 PRINT L$(S1, DB(S1)): PRINT
1000 IF S1 <= 2 THEN PRINT "WHERE DUCT SILENCERS ARE USED, THE DUCT CROSS-SECTION AREA MUST BE COMPRISED"
1010 IF S1 <= 2 THEN PRINT "OF ONE OR MORE OF THE FOLLOWING SILENCER SELECTIONS": PRINT
1020 IF S1 <= 2 THEN I1 = DB(S1) - 3 ELSE I1 = DB(S1) - 2: 'Number of inputs for silencer dimensions
1030 '
1040 ' Print out duct silencer size menu
1050 '
1060 FOR I = 1 TO I1 STEP 2
1070 IF S1 <= 2 THEN PRINT LEFT$(L$(S1, I), 11); TAB(20); LEFT$(L$(S1, I + 1), 11)
1080 IF S1 > 2 THEN NA = LEN(L$(S1, I + 1)): IF NA <= 18 THEN NL = 13 ELSE NL = 14
1090 IF S1 > 2 THEN PRINT LEFT$(L$(S1, I), NL); TAB(30); LEFT$(L$(S1, I + 1), NL)
1100 NEXT I
1110 INPUT "SELECTION: ", S3
1120 IF S3 >= 1 AND S3 <= I1 THEN GOTO 1130 ELSE GOTO 1110
1130 '
1140 IF S1 <= 2 THEN NN = LEN(L$(S1, S3)) - 11: AR = VAL(RIGHT$(L$(S1, S3), NN))
1150 IF S1 > 2 THEN AR = VAL(RIGHT$(L$(S1, S3), 4)): DIA$ = LEFT$(L$(S1, S3), 6): DIA = VAL(RIGHT$(DIA$, 2))
1160 CLS : LOCATE 3, 1: PRINT "SINGLE SILENCER FACE AREA = "; AR; " SQ IN"
```

```
1170 IF S1 <= 2 THEN PRINT "SILENCER LENGTH = 7 FT"
1180 IF S1 <= 2 THEN LOCATE 6, 1: INPUT "NO OF SILENCERS NECESSARY TO MAKE UP TOTAL DUCT FACE
AREA: ", NS
1190 IF S1 > 2 THEN PRINT "SILENCER DIAMETER = "; DIA; " IN": NS = 1
1200 AR = NS * AR / 144: PRINT : INPUT "ENTER DUCT FLOW VOLUME (CFM):", FV
1210 VL = CINT(FV / AR): VA = VL
1220 PRINT : PRINT "SILENCER FACE VELOCITY = "; VL; " FT/MIN"
1230 '
1240 IF S1 <= 2 THEN PD = (SP(S1, 1) * 7 ^ SP(S1, 2)) * (SP(S1, 3) * VL ^ SP(S1, 4))
1250 '
1260 ' Equation 5.25
1270 '
1280 IF S1 > 2 THEN PDC = SP(S1, 1) * DIA ^ SP(S1, 2): PD = SP(S1, 3) * (FV / PDC) ^ SP(S1, 4)
1290 '
1300 ' Equation 5.26
1310 '
1320 PD = CINT(PD * 100) / 100
1330 PRINT "SILENCER PRESS. DROP = "; PD; " IN H2O"
1340 PRINT : PRINT "ARE THERE SYSTEM COMPONENTS UPSTREAM FROM THE SILENCER?   (1. YES   2. NO)"
1350 INPUT "SELECTION: ", S4
1360 IF S4 < 1 OR S4 > 2 THEN GOTO 1350 ELSE GOTO 1370
1370 IF S4 = 2 THEN CUP = 1: GOTO 1420
1380 INPUT "DISTANCE TO UPSTREAM SYSTEM COMPONENT (FT): ", UPDIST
1390 DEQ = SQR(4 * AR / 3.1415)
1400 NDEQ = UPDIST / DEQ
1410 GOSUB 2120: CUP = C
1420 PRINT : PRINT "ARE THERE SYSTEM COMPONENTS DOWNSTREAM FROM THE SILENCER?   (1. YES   2. NO)"
1430 INPUT "SELECTION: ", S5
1440 IF S5 < 1 OR S5 > 2 THEN GOTO 1430 ELSE GOTO 1450
1450 IF S5 = 2 THEN CDOWN = 1: GOTO 1500
1460 INPUT "DISTANCE TO DOWNSTREAM SYSTEM COMPONENT (FT): ", DOWNDIST
1470 DEQ = SQR(4 * AR / 3.1415)
1480 NDEQ = DOWNDIST / DEQ
1490 GOSUB 2120: CDOWN = C
1500 C5 = CUP * CDOWN
1510 PD = PD * C5: PD = CINT(PD * 100) / 100
1520 PRINT : PRINT "SILENCER PRESS. DROP W SYSTEM COMPONENTS = "; PD; " IN H2O"
1530 PRINT : PRINT "SOUND PROPAGATES": PRINT "1. WITH AIR FLOW"; SPC(10); "2. AGAINST AIR FLOW"
1540 INPUT "SELECTION: ", S6
1550 IF S6 >= 1 AND S6 <= 2 THEN GOTO 1560 ELSE GOTO 1540
1560 PRINT
1570 PRINT "CONTINUE? (Y OR N)"
1580 Q$ = INKEY$: IF Q$ = "Y" OR Q$ = "N" OR Q$ = "y" OR Q$ = "n" THEN GOTO 1590 ELSE GOTO 1580
1590 IF Q$ = "N" OR Q$ = "n" THEN FLAG = 2: GOTO 580
1600 VL$ = STR$(VL)
1610 IF S1 <= 2 THEN GOSUB 2480: ' For rectangular silencers
1620 IF S1 > 2 THEN GOSUB 2650: ' For circular silencers
1630 '
1640 GOSUB 2200: ' Interpolation subroutine
1650 '
1660 ' Face area adjustment factor
1670 '
1680 IF S1 <= 2 THEN FAAF = 10 * LOG(AR) / LOG(10) - 6: ' equation 5.22
1690 IF S1 > 2 THEN FAAF = 10 * LOG(AR) / LOG(10) - 4.755: ' equation 5.23
1700 '
1710 ' Face area adjustment factor is added to Air flow generated noise
1720 '
1730 FOR I = 1 TO 8
1740   AFGNF(I) = AFGNF(I) + FAAF: ' Lw3, equation 5.31
1750 NEXT I
1760 '
1770 ' Calculation of Sound Power Level After The Duct Silencer
1780 '
1790 FOR I = 1 TO 8
1800   LW2 = PWL(I) - DILF(I): ' equation 5.30
1810   T%(I) = 10 * LOG(10 ^ (LW2 / 10) + 10 ^ (AFGNF(I) / 10)) / LOG(10): ' eq. 5.32
1820 NEXT I
1830 '
1840 ' Output statements
1850 '
1860 PRINT
1870 T1$ = "SOUND POWER LEVEL AFTER SIL., dB"
1880 L2$ = "ORIGINAL SOUND POWER LEVEL, dB"
1890 L3$ = "DUCT INSERTION LOSS, dB"
1900 L4$ = "AIR FLOW GENERATED NOISE, dB"
```

```
1910 A$ = STRING$(41, CHR$(45)) + "  ___ ___ ___ ___ ___ ___ ___ ___"
1920 PRINT SPC(45); "1/1 OCT BAND CENTER FREQ"
1930 PRINT SPC(10); "DESCRIPTION"; SPC(21); " 63 125 250 500  1K  2K  4K  8K"
1940 PRINT A$
1950 PRINT USING "\                                        \"; L2$;
1960 PRINT USING "  ###"; PWL(1); PWL(2); PWL(3); PWL(4); PWL(5); PWL(6); PWL(7); PWL(8)
1970 PRINT USING "\                                        \"; L3$;
1980 PRINT USING "  ###"; -DILF(1); -DILF(2); -DILF(3); -DILF(4); -DILF(5); -DILF(6); -DILF(7); -DILF(8)
1990 PRINT A$
2000 FOR I = 1 TO 8: TEMP(I) = PWL(I) - DILF(I): NEXT I
2010 PRINT USING "\                                        \"; "SILENCER EXIT Lw, dB";
2020 PRINT USING "  ###"; TEMP(1); TEMP(2); TEMP(3); TEMP(4); TEMP(5); TEMP(6); TEMP(7); TEMP(8)
2030 PRINT USING "\                                        \"; L4$;
2040 PRINT USING "  ###"; AFGNF(1); AFGNF(2); AFGNF(3); AFGNF(4); AFGNF(5); AFGNF(6); AFGNF(7); AFGNF(8)
2050 PRINT STRING$(41, CHR$(45)) + "  ___ ___ ___ ___ ___ ___ ___ ___"
2060 PRINT USING "\                                        \"; T1$;
2070 PRINT USING "  ###"; T%(1); T%(2); T%(3); T%(4); T%(5); T%(6); T%(7); T%(8)
2080 END
2090 '
2100 'Coefficient for system component effect of pressure drop
2110 '
2120 IF NDEQ > 3 THEN C = 1: GOTO 2150
2130 IF NDEQ > 1 AND NDEQ <= 3 THEN C = CINT(10 * (2.33 - .45 * NDEQ)) / 10
2140 IF NDEQ >= 0 AND NDEQ <= 1 THEN C = CINT(10 * (4.02 - 2.1 * NDEQ)) / 10
2150 RETURN
2160 '
2170 ' Interpolating subroutine for flow velcities. Matches flow velcity to
2180 ' corresponding DIL and AFGN
2190 '
2200 ADJ = 0
2210 IF VA <= FV(S1, 1) THEN I3 = 1: GOTO 2280
2220 IF VA >= FV(S1, DD(S1)) THEN I3 = DD(S1): GOTO 2280
2230 FOR I = 1 TO DD(S1)
2240     I3 = I
2250     IF FV(S1, I) <= VA AND FV(S1, I + 1) > VA THEN GOTO 2270
2260 NEXT I
2270 ADJ = (VA - FV(S1, I)) / (FV(S1, I + 1) - FV(S1, I))
2280 FOR J = 1 TO 8: DILF(J) = DIL(I3, J) + ADJ * (DIL(1 + I3, J) - DIL(I3, J)): NEXT J
2290 ' aiflow generated noise
2300 '
2310 '
2320 ADJ = 0
2330 IF VA <= FV(S1, DD(S1) + 1) THEN I4 = 1: GOTO 2400
2340 IF VA >= FV(S1, DA(S1)) THEN I4 = DA(S1) - DD(S1): GOTO 2400
2350 FOR I = (DD(S1) + 1) TO DA(S1)
2360     I4 = I - DD(S1)
2370     IF FV(S1, I) <= VA AND FV(S1, I + 1) > VA THEN GOTO 2390
2380 NEXT I
2390 ADJ = (VA - FV(S1, I)) / (FV(S1, I + 1) - FV(S1, I))
2400 FOR J = 1 TO 8: AFGNF(J) = AFGN(I4, J) + ADJ * (AFGN(1 + I4, J) - AFGN(I4, J)): NEXT J
2410 RETURN
2420 '
2430 ' N1 is for DIL values, N2 is for Airflow generated noise
2440 '
2450 '
2460 ' rectangular ducts
2470 '
2480 IF S6 = 1 THEN N1 = 0: N2 = DD(S1): ' For sound propagating with airflow
2490 IF S6 = 2 THEN N1 = DD(S1) + 5: N2 = 2 * DD(S1) + 5: ' For sound propagating against airflo
2500 FOR I = 1 TO DD(S1)
2510 FOR J = 1 TO 8
2520 DIL(I, J) = D(S1, N1 + I, J): 'Duct insertion loss for all available velocities
2530 NEXT J
2540 NEXT I
2550 FOR I = 1 TO 5
2560 FOR J = 1 TO 8
2570 AFGN(I, J) = D(S1, N2 + I, J): ' Airflow generated noise for all available velocities
2580 NEXT J
2590 NEXT I
2600 '
2610 RETURN
2620 '
2630 ' Circular ducts
```

```
2640 '
2650 IF S6 = 1 THEN N1 = 0: N2 = 3
2660 IF S6 = 2 THEN N1 = 6: N2 = 9
2670 FOR I = 1 TO 3
2680 FOR J = 1 TO 8
2690 DIL(I, J) = D(S1, N1 + I, J): 'Duct insertion loss for all available velocities
2700 AFGN(I, J) = D(S1, N2 + I, J): ' Airflow generated noise for all available velocities
2710 NEXT J
2720 NEXT I
2730 RETURN
2740 '
2750 DATA 5,15,10,3: ' For Rectangular, High Pressure Drop Duct Silencers
2760 DATA 6,13,12,3: ' For Rectangular, Low Pressure Drop Duct Silencer
2770 DATA 6,20,12,3: ' For Circular, High Pressure Drop Silencers
2780 DATA 6,20,12,3: ' For Circular, Low Pressure Drop Silencers
2790 '
2800 'Rectangular, High Pressure Drop Duct Silencers
2810 '
2820 DATA 5.,14.,31.,45.,51.,53.,51.,33.: '7 ft   +1000 fpm
2830 DATA 4.,12.,26.,43.,47.,48.,47.,30.: '7 ft   +2000 fpm
2840 DATA 58.,52.,42.,36.,37.,35.,30.,30.: 'air flow gen. ns. +1000 fpm
2850 DATA 64.,57.,58.,49.,45.,49.,48.,46.: 'air flow gen. ns. +1500 fpm
2860 DATA 72.,65.,64.,63.,55.,56.,57.,56.: 'air flow gen. ns. +2000 fpm
2870 DATA 7.,19.,36.,44.,48.,50.,48.,29.: '7 ft   -1000 fpm
2880 DATA 4.,19.,34.,42.,46.,48.,46.,28.: '7 ft   -2000 fpm
2890 DATA 69.,66.,65.,75.,71.,73.,79.,76.: 'air flow gen. ns. -2000 fpm
2900 DATA 55.,52.,54.,55.,55.,64.,64.,55.: 'air flow gen. ns. -1000 fpm
2910 DATA 61.,58.,59.,61.,61.,66.,75.,66.: 'air flow gen. ns. -1500 fpm
2920 '
2930 'Rectangular, Low Pressure Drop Duct Silencer
2940 '
2950 DATA 2.,8.,18.,33.,41.,47.,24.,15.: '7 ft   +1000 fpm
2960 DATA 1.,8.,17.,32.,39.,44.,24.,16.: '7 ft   +2000 fpm
2970 DATA 1.,7.,17.,30.,37.,42.,24.,15.: '7 ft   +2500 fpm
2980 DATA 58.,51.,40.,34.,35.,28.,27.,19.: 'air flow gen. ns. +1000 fpm
2990 DATA 67.,61.,58.,53.,51.,54.,52.,45.: 'air flow gen. ns. +2000 fpm
3000 DATA 74.,68.,65.,62.,56.,59.,60.,55.: 'air flow gen. ns. +2500 fpm
3010 DATA 1.,11.,21.,35.,41.,45.,22.,12.: '7 ft   -1000 fpm
3020 DATA 1.,11.,21.,36.,40.,42.,21.,11.: '7 ft   -2000 fpm
3030 DATA 1.,9.,21.,34.,38.,40.,20.,10.: '7 ft   -2500 fpm
3040 DATA 58.,49.,46.,44.,49.,45.,34.,25.: 'air flow gen. ns. -1000 fpm
3050 DATA 70.,61.,59.,56.,57.,62.,58.,50.: 'air flow gen. ns. -2000 fpm
3060 DATA 74.,67.,64.,62.,61.,65.,65.,57.: 'air flow gen. ns. -2500 fpm
3070 '
3080 'Circlar, High Pressure Drop Duct Silencer
3090 '
3100 DATA 4.,7.,21.,32.,38.,38.,26.,20.: '24" d.,4' long, +1000 fpm
3110 DATA 4.,7.,20.,32.,38.,38.,27.,21.: '24" d.,4' long, +2000 fpm
3120 DATA 4.,7.,19.,31.,38.,38.,27.,21.: '24" d.,4' long, +3000 fpm
3130 DATA 62.,43.,38.,39.,36.,25.,22.,28.: 'air flow gen. ns. +1000 fpm
3140 DATA 62.,58.,53.,54.,53.,51.,45.,35.: 'air flow gen. ns. +2000 fpm
3150 DATA 72.,66.,62.,64.,63.,64.,61.,54.: 'air flow gen. ns. +3000 fpm
3160 DATA 5.,9.,23.,33.,39.,37.,25.,19.: '24" d.,4' long, -1000 fpm
3170 DATA 5.,9.,23.,34.,37.,36.,24.,16.: '24" d.,4' long, -2000 fpm
3180 DATA 6.,10.,24.,34.,37.,36.,24.,16.: '24" d.,4' long, -3000 fpm
3190 DATA 56.,43.,41.,38.,37.,31.,23.,28.: 'air flow gen. ns. -1000 fpm
3200 DATA 66.,56.,54.,54.,57.,54.,49.,41.: 'air flow gen. ns. -2000 fpm
3210 DATA 76.,64.,63.,64.,67.,67.,65.,60.: 'air flow gen. ns. -3000 fpm
3220 '
3230 'Circlar, Low Pressure Drop Duct Silencer
3240 '
3250 DATA 4.,6.,13.,26.,32.,24.,16.,14.: '24" d.,4' long, +1000 fpm
3260 DATA 4.,5.,13.,25.,32.,24.,16.,13.: '24" d.,4' long, +2000 fpm
3270 DATA 4.,6.,13.,23.,31.,24.,16.,13.: '24" d.,4' long, +3000 fpm
3280 DATA 60.,44.,39.,34.,29.,25.,24.,30.: 'air flow gen. ns. +1000 fpm
3290 DATA 62.,58.,48.,47.,49.,45.,38.,31.: 'air flow gen. ns. +2000 fpm
3300 DATA 69.,63.,55.,55.,57.,58.,54.,47.: 'air flow gen. ns. +3000 fpm
3310 DATA 5.,7.,16.,28.,35.,25.,17.,15.: '24" d.,4' long, -1000 fpm
3320 DATA 4.,7.,16.,28.,35.,25.,17.,15.: '24" d.,4' long, -2000 fpm
3330 DATA 6.,7.,16.,29.,35.,25.,16.,15.: '24" d.,4' long, -3000 fpm
3340 DATA 58.,46.,43.,38.,33.,30.,24.,30.: 'air flow gen. ns. -1000 fpm
3350 DATA 65.,52.,50.,49.,48.,44.,36.,33.: 'air flow gen. ns. -2000 fpm
3360 DATA 72.,57.,57.,57.,56.,57.,52.,49.: 'air flow gen. ns. -3000 fpm
3370 '
3380 ' Labels for Rectangular Standard Pressure Drop Silencers
```

```
3390 '
3400 DATA "1.   6X12     72"
3410 DATA "2.  12X12    144"
3420 DATA "3.  18X12    216"
3430 DATA "4.  24X12    288"
3440 DATA "5.  30X12    360"
3450 DATA "6.  36X12    432"
3460 DATA "7.   6X24    144"
3470 DATA "8.  18X24    432"
3480 DATA "9.  24X24    576"
3490 DATA "10. 30X24    720"
3500 DATA "11. 36x24    864"
3510 DATA "12. 36x36   1296"
3520 DATA "HIGH PRESSURE DROP"
3530 DATA "HIGH PRESSURE DROP:  SILENCER LENGTH (ft)"
3540 DATA "HIGH PRESSURE DROP:  NOMINAL FACE SIZE W X H (IN x IN)"
3550 '
3560 'Labels for Rectangular Low Pressure Drop Silencers
3570 '
3580 DATA "1.  7.5X12    90"
3590 DATA "2.  15X12    180"
3600 DATA "3.  15X18    270"
3610 DATA "4.  15X24    360"
3620 DATA "5.  15X30    450"
3630 DATA "6.  15X36    540"
3640 DATA "7.  30X24    720"
3650 DATA "8.  30X30    900"
3660 DATA "9.  30x36   1080"
3670 DATA " "
3680 DATA "LOW PRESSURE DROP"
3690 DATA "LOW PRESSURE DROP:  SILENCER LENGTH (FT)"
3700 DATA "LOW PRESSURE DROP:  NOMINAL FACE SIZE W X H (IN x IN)"
3710 '
3720 ' Labels for Circular Standard Pressure Drop Silencers
3730 '
3740 DATA "1.  12    36    113"
3750 DATA "2.  14    36    154"
3760 DATA "3.  16    36    202"
3770 DATA "4.  18    36    255"
3780 DATA "5.  20    40    314"
3790 DATA "6.  22    44    380"
3800 DATA "7.  24    48    452"
3810 DATA "8.  26    52    531"
3820 DATA "9.  28    56     61"
3830 DATA "10. 30    60    707"
3840 DATA "11. 32    64    805"
3850 DATA "12. 36    72   1018"
3860 DATA "13. 40    80   1257"
3870 DATA "14. 44    88   1521"
3880 DATA "15. 48    96   1810"
3890 DATA "16. 52   104   2124"
3900 DATA "17. 56   112   2462"
3910 DATA "18. 60   120   2827"
3920 DATA "HIGH PRESSURE DROP, INSULATED CONE AND OUTER SURFACE: SILENCER LENGTH (FT)"
3930 DATA "HIGH PRESSURE DROP, INSULATED CONE AND OUTER SURFACE: NOMINAL FACE SIZE DIA X L (IN x IN)"
3940 '
3950 ' Labels for Circular Low Pressure Drop Silencers
3960 '
3970 DATA "1.  12    36     113"
3980 DATA "2.  14    36     154"
3990 DATA "3.  16    36     202"
4000 DATA "4.  18    36     255"
4010 DATA "5.  20    40     314"
4020 DATA "6.  22    44     380"
4030 DATA "7.  24    48     452"
4040 DATA "8.  26    52     531"
4050 DATA "9.  28    56     616"
4060 DATA "10. 30    60     707"
4070 DATA "11. 32    64     805"
4080 DATA "12. 36    72    1018"
4090 DATA "13. 40    80    1257"
4100 DATA "14. 44    88    1521"
4110 DATA "15. 48    96    1810"
4120 DATA "16. 52   104    2124"
```

```
4130 DATA "17. 56    112      2462"
4140 DATA "18. 60    120      2827"
4150 DATA "LOW PRESSURE DROP, INSULATED CONE AND OUTER SURFACE: SILENCER LENGTH (FT)"
4160 DATA "LOW PRESSURE DROP, INSULATED CONE AND OUTER SURFACE: NOMINAL FACE SIZE DIA X L (IN x IN)"
4170 '
4180 DATA .6464418,.3971121,2.636889E-07,2.011962: ' Coefficients for pressure drop
4190 DATA .6015026,.4626969,9.802441E-08,2.010546: ' Regression
4200 DATA 5.107506E-03,1.999595,4.007338E-08,2.001673
4210 DATA 5.097187E-03,2.000361,1.104126E-08,2.021844
4220 '
4230 DATA 1000,2000,1000,1500,2000: 'Flow velocities Corresponding
4240 DATA 1000,2000,2500,1000,2000,2500: ' to experimental
4250 DATA 1000,2000,3000,1000,2000,3000: ' data
4260 DATA 1000,2000,3000,1000,2000,3000
```

SAMPLE OUTPUT FOR EXAMPLE 5.9

SOUND ATTENUATION AND REGENERATED NOISE OF DUCT SILENCERS

INPUT SOUND POWER LEVELS IN DUCT BEFORE SILENCER
SOUND POWER LEVEL FOR THE 62.5 HZ OCTAVE BAND ? 91
SOUND POWER LEVEL FOR THE 125 HZ OCTAVE BAND ? 87
SOUND POWER LEVEL FOR THE 250 HZ OCTAVE BAND ? 83
SOUND POWER LEVEL FOR THE 500 HZ OCTAVE BAND ? 82
SOUND POWER LEVEL FOR THE 1000 HZ OCTAVE BAND ? 78
SOUND POWER LEVEL FOR THE 2000 HZ OCTAVE BAND ? 76
SOUND POWER LEVEL FOR THE 4000 HZ OCTAVE BAND ? 72
SOUND POWER LEVEL FOR THE 8000 HZ OCTAVE BAND ? 70

CONTINUE? (Y OR N)

RECTANGULAR DUCT SILENCERS
1. STANDARD PRESSURE DROP
2. LOW PRESSURE DROP

CIRCULAR DUCT SILENCERS, INSULATED CONE AND OUTER SURFACE
3. STANDARD PRESSURE DROP
4. LOW PRESSURE DROP

SELECTION: 2

STANDARD LOW PRESSURE DROP: NOMINAL FACE SIZE W X H (IN x IN)

WHERE DUCT SILENCERS ARE USED, THE DUCT CROSS-SECTION AREA MUST BE COMPRISED
OF ONE OR MORE OF THE FOLLOWING SILENCER SELECTIONS

1. 7.5X12   2. 15X12
3. 15X18   4. 15X24
5. 15X30   6. 15X36
7. 30X24   8. 30X30
9. 30x36
SELECTION: 7

SINGLE SILENCER FACE AREA = 720 SQ IN
SILENCER LENGTH = 7 FT

NO OF SILENCERS NECESSARY TO MAKE UP TOTAL DUCT FACE AREA: 1

ENTER DUCT FLOW VOLUME (CFM):10000

SILENCER FACE VELOCITY = 2000 FT/MIN
SILENCER PRESS. DROP = .63 IN H2O

ARE THERE SYSTEM COMPONENTS UPSTREAM FROM THE SILENCER? (1. YES 2. NO)
SELECTION: 2

ARE THERE SYSTEM COMPONENTS DOWNSTREAM FROM THE SILENCER? (1. YES 2. NO)
SELECTION: 1
DISTANCE TO DOWNSTREAM SYSTEM COMPONENT (FT): 5

SILENCER PRESS. DROP W SYSTEM COMPONENTS = .88 IN H2O

SOUND PROPAGATES
1. WITH AIR FLOW   2. AGAINST AIR FLOW
SELECTION: 1

CONTINUE? (Y OR N)

DESCRIPTION	63	125	250	500	1K	2K	4K	8K
ORIGINAL SOUND POWER LEVEL, dB	91	87	83	82	78	76	72	70
DUCT INSERTION LOSS, dB	−1	−8	−17	−32	−39	−44	−24	−16
SILENCER EXIT Lw, dB	90	79	66	50	39	32	48	54
AIR FLOW GENERATED NOISE, dB	68	62	59	54	52	55	53	46
SOUND POWER LEVEL AFTER SIL., dB	90	79	67	55	52	55	54	55

(1/1 OCT BAND CENTER FREQ)

DUCT BRANCH SOUND POWER DIVISION

PROGRAM OUTLINE

Input system parameters:
    Main duct dimensions
    Number of branch ducts
    Branch duct dimensions
    Select duct branch for duct branch attenuation
Calculate:
    Cross–sectional areas
    Cut–off frequency, f_{co}
    Attenuation of sound power level, ΔL_{B_i}

COMPUTER PROGRAM – BRANCH.BAS

```
10  ' >DUCT BRANCH SOUND POWER DIVISION COMPUTER PROGRAM<
20  '
30  DIM T%(8), S2(5), DWB(5), DHB(5)
40  PI = 3.141593
50  '
60  '    LIST OF VARIABLES
70  '
80  '    S1 = selection for geometry of main feeder duct
90  '    S2 = selection for geometry of branch duct
100 '     DWM = smaller main rectangular feeder dimension (in.)
110 '     DHM = larger main rectangular feeder dimension (in.)
120 '     DDM = main circular feeder duct diameter (in.)
130 '     NB = number of branch ducts
140 '     SB = sum of branch duct cross sectional area (sq in)
150 '     DWB() = smaller branch rectangular duct dimension (in)
160 '     DHB() = larger branch duct dimension (in.)
170 '     DDB() = branch circular duct diameter (in)
180 '     BN = branch number which sound travels down
190 '     MA = main duct cross sectional area
200 '     CA = cross  sectional area of branch which sound travels down
210 '     DL1, DL2 = duct branch sound power division (dB)
220 '     FCO = Cutoff frequency (Hz)
230 '     T%() = Duct branch sound power division output
240 '
250 CLS : PRINT "SOUND ATTENUATION ASSOCIATED WITH A DUCT BRANCH POWER DIVISION"
260 '
270 ' Enter main feeder duct dimensions
280 '
290 PRINT : INPUT "MAIN FEEDER DUCT (1. RECTANGULAR    2. CIRCULAR): ", S1
300 IF S1 >= 1 AND S1 <= 2 THEN GOTO 310 ELSE GOTO 290
310 '
320 IF S1 = 2 THEN GOTO 360
330 INPUT " ENTER SMALLER MAIN FEEDER DUCT DIMENSION: ", DWM
340 INPUT " ENTER LARGER MAIN FEEDER DUCT DIMENSION: ", DHM
350 GOTO 370
360 INPUT " ENTER MAIN FEEDER DUCT DIAMETER (IN): ", DDM
370 INPUT "ENTER NUMBER OF BRANCH DUCTS: (1 <= NUMBER <= 5): ", NB
380 PRINT
390 SB = 0: ' Initialize area of branch ducts to 0
400 '
410 ' Enter branch duct dimensions
420 '
430 FOR I = 1 TO NB
440    IF I = 1 THEN PRINT "DUCT 1 IS ALWAYS THE CONTINUATION OF THE MAIN FEEDER DUCT."
450    IF I > 1 THEN PRINT "DUCT "; I; " IS A BRANCH DUCT OFF OF THE MAIN FEEDER DUCT."
460    PRINT "DUCT NO. "; I; : INPUT " (1. RECTANGULAR    2. CIRCULAR): ", S2(I)
470    IF S2(I) >= 1 AND S2(I) <= 2 THEN GOTO 480 ELSE GOTO 460
480 '
490    IF S2(I) = 1 THEN GOTO 500 ELSE GOTO 540: ' For rectangular ducts
500    INPUT " ENTER SMALLER BRANCH DUCT DIMENSION (IN): ", DWB(I)
510    INPUT " ENTER LARGER BRANCH DUCT DIMENSION (IN): ", DHB(I)
```

```
520 ,   SB = SB + DWB(I) * DHB(I)
530 ,
540     IF S2(I) = 2 THEN GOTO 550 ELSE GOTO 570: ' For circular ducts
550     INPUT " ENTER BRANCH DUCT DIAMETER (IN): ", DDB(I)
560     SB = SB + PI * DDB(I) ^ 2 / 4
570 NEXT I
580 PRINT
590 '
600 PRINT "FOLLOW SOUND DOWN DUCT BRANCH NUMBER: (1 <= NUMBER <="; NB; : INPUT "): ", BN
610 IF BN >= 1 AND BN <= NB THEN GOTO 620 ELSE GOTO 600
620 PRINT
630 PRINT "CONTINUE? (Y OR N)"
640 Q$ = INKEY$: IF Q$ = "Y" OR Q$ = "N" OR Q$ = "y" OR Q$ = "n" THEN GOTO 650 ELSE GOTO 640
650 IF Q$ = "N" OR Q$ = "n" THEN GOTO 250
660 '
670 ' Area of main feeder duct
680 '
690 IF S1 = 1 THEN MA = DWM * DHM
700 IF S1 = 2 THEN MA = PI * DDM ^ 2 / 4
710 '
720 ' Area of branch which sound travels down
730 '
740 IF S2(NB) = 1 THEN CA = DWB(BN) * DHB(BN)
750 IF S2(NB) = 2 THEN CA = PI * DDB(BN) ^ 2 / 4
760 '
770 ' Cutoff frequency
780 '
790 IF S1 = 1 THEN FCO = 1125 * 12 / (2 * DHM): ' Equation 5.34
800 IF S1 = 2 THEN FCO = .586 * 1125 * 12 / DDM: ' Equation 5.35
810 '
820 DL1 = 10 * LOG(CA / SB) / LOG(10): ' Equation 5.33
830 DL2 = 10 * LOG(1 - ((SB / MA - 1) / (SB / MA + 1)) ^ 2) / LOG(10): ' Eq. 5.33
840 FOR I = 1 TO 8
850 FR = 62.5 * 2 ^ (I - 1): 'Frequency
860 IF FR < FCO THEN T%(I) = -DL1 - DL2
870 IF FR >= FCO THEN T%(I) = -DL1
880 NEXT I
890 '
900 ' Output statements
910 '
920 BN$ = STR$(BN)
930 T1$ = "BRANCH" + BN$ + " LW DIV. ATTENUATION, dB"
940 PRINT
950 PRINT SPC(45); "1/1 OCT BAND CENTER FREQ"
960 PRINT SPC(10); "DESCRIPTION"; SPC(21); " 63  125  250  500  1K   2K   4K   8K"
970 PRINT STRING$(41, CHR$(45)) + " ___ ___ ___ ___ ___ ___ ___ ___"
980 PRINT USING "\                                     \"; T1$;
990 PRINT USING " ###"; T%(1); T%(2); T%(3); T%(4); T%(5); T%(6); T%(7); T%(8)
1000 END
```

SAMPLE OUTPUT FOR EXAMPLE 5.10

SOUND ATTENUATION ASSOCIATED WITH A DUCT BRANCH SOUND POWER DIVISION

MAIN FEEDER DUCT (1. RECTANGULAR   2. CIRCULAR): 2
    ENTER MAIN FEEDER DUCT DIAMETER (IN): 18
ENTER NUMBER OF BRANCH DUCTS (1 <= NUMBER <= 5): 2

DUCT 1 IS ALWAYS THE CONTINUATION OF THE MAIN FEEDER DUCT.
DUCT NO. 1 (1. RECTANGULAR   2. CIRCULAR): 2
    ENTER BRANCH DUCT DIAMETER (IN): 12
DUCT 2 IS A BRANCH DUCT OFF OF THE MAIN FEEDER DUCT.
DUCT NO. 2 (1. RECTANGULAR   2. CIRCULAR): 2
    ENTER BRANCH DUCT DIAMETER (IN): 6

FOLLOW SOUND DOWN DUCT BRANCH NUMBER (1 <= NUMBER <= 2): 2

CONTINUE? (Y OR N)

DESCRIPTION	63	125	250	500	1K	2K	4K	8K
BRANCH 2 LW DIV. ATTENUATION, dB	7	7	7	7	7	7	7	7

1/1 OCT BAND CENTER FREQ

DUCT END REFLECTION LOSS

PROGRAM OUTLINE

Input system parameters:
- Diffuser dimensions
- Diffuse termination configuration

Calculate
- ΔL, sound attenuation associated with the end reflection loss

COMPUTER PROGRAM – ENDRLOSS.BAS

```
10  ' >DUCT ENE REFLECTION LOSS COMPUTER PROGRAM<
20  '
30  DIM T%(8)
40  PI = 3.141593
50  '
60  '       LIST OF VARIABLES
70  '
80  '       DDM = diffuser diameter (in.)
90  '       DHM = sarger rectangular dimension (in.)
100 '       DWM = smaller rectangular dimension (in.)
110 '       S1  = selection for geometry of diffuser
120 '       S2  = selection for type of diffuser termination
130 '
140 CLS : PRINT "DUCT END REFLECTION LOSS": PRINT
150 INPUT "TYPE DIFFUSER (1. RECTANGULAR    2. CIRCULAR): ", S1
160 IF S1 >= 1 AND S1 <= 2 THEN GOTO 170 ELSE GOTO 150
170 IF S1 = 2 THEN GOTO 240
180 '
190 ' Enter diffuser dimensions
200 '
210 INPUT "ENTER SMALLER DIFFUSER DIMENSION (IN): ", DWM
220 INPUT "ENTER LARGER DIFFUSER DIMENSION (IN): ", DHM
230 GOTO 250
240 INPUT "ENTER DIFFUSER DIAMETER (IN): ", DDM
250 PRINT : INPUT "DIFFUSER TERMINATION (1. FREE SPACE    2. FLUSH WITH WALL): ", S2
260 IF S2 >= 1 AND S2 <= 2 THEN GOTO 270 ELSE GOTO 250
270 PRINT
280 PRINT "CONTINUE? (Y OR N)"
290 Q$ = INKEY$: IF Q$ = "Y" OR Q$ = "N" OR Q$ = "y" OR Q$ = "n" THEN GOTO 300 ELSE GOTO 290
300 IF Q$ = "N" OR Q$ = "n" THEN GOTO 140
310 '
320 ' Calculate effective diameter for rectangular diffuser
330 '
340 IF S1 = 1 THEN DDM = SQR(4 * DWM * DHM / PI)
350 '
360 ' Calculate end reflection loss
370 '
380 FOR I = 1 TO 8
390 FR = 62.5 * 2 ^ (I - 1): ' Frequency (Hz)
400 '     Equation 5.37
410 IF S2 = 1 THEN T%(I) = 10 * LOG(1 + (1125 * 12 / (PI * FR * DDM)) ^ 1.88) / LOG(10)
420 '     Equation 5.38
430 IF S2 = 2 THEN T%(I) = 10 * LOG(1 + (.8 * 1125 * 12 / (PI * FR * DDM)) ^ 1.88) / LOG(10)
440 NEXT I
450 '
460 ' Output statements
470 '
480 PRINT
490 PRINT SPC(45); "1/1 OCT BAND CENTER FREQ"
500 PRINT SPC(10); "DESCRIPTION"; SPC(21); " 63  125  250  500  1K   2K   4K   8K"
510 PRINT STRING$(41, CHR$(45)) + " ___ ___ ___ ___ ___ ___ ___ ___"
520 PRINT USING "\                                        \"; "DUCT END REFLECTION SOUND ATTENUATION, dB";
530 PRINT USING " ###"; T%(1); T%(2); T%(3); T%(4); T%(5); T%(6); T%(7); T%(8)
540 END
```

SAMPLE OUTPUT

DUCT END REFLECTION LOSS

TYPE DIFFUSER (1. RECTANGULAR   2. CIRCULAR): 2
ENTER DIFFUSER DIAMETER (IN): 12

DIFFUSER TERMINATION (1. FREE SPACE   2. FLUSH WITH WALL): 1

CONTINUE? (Y OR N)

DESCRIPTION	63	125	250	500	1K	2K	4K	8K
DUCT END REFLECTION SOUND ATT., dB	14	9	5	2	1	0	0	0

(1/1 OCT BAND CENTER FREQ)

TERMINAL VOLUME REGULATION UNITS

COMPUTER PROGRAM – VOLREG.BAS

```
10 '>SOUND ATTENUATION—TERMINAL VOLUME REGULATION UNITS COMPUTER PROGRAM<
20 DIM T%(8)
30 T%(1) = 0: T%(2) = 5: T%(3) = 10: 'T% is Attenuation, see Table 4.25
40 FOR I = 4 TO 8
50 T%(I) = 15
60 NEXT I
70 PRINT SPC(45); "1/1 OCT BAND CENTER FREQ"
80 PRINT SPC(10); "DESCRIPTION"; SPC(21); " 63 125 250 500  1K   2K   4K   8K"
90 PRINT STRING$(41, CHR$(45)) + " ___ ___ ___ ___ ___ ___ ___ ___"
100 PRINT USING "\                                                    \"; "TERM VOL REG UNIT SOUND ATTENUATION, dB";
110 PRINT USING " ###"; T%(1); T%(2); T%(3); T%(4); T%(5); T%(6); T%(7); T%(8)
120 END
```

SAMPLE OUTPUT

		1/1 OCT BAND CENTER FREQ						
DESCRIPTION	63	125	250	500	1K	2K	4K	8K
TERM VOL REG UNIT SOUND ATTEN., dB	0	5	10	15	15	15	15	15

COMPUTER PROGRAMS FOR CHAPTER 6

APPENDIX 5

COMPUTER PROGRAMS FOR CHAPTER 6

RECTANGULAR DUCT BREAKOUT AND BREAKIN

PROGRAM OUTLINE

Read in mass/unit area's for sheet metal
Input system parameters
    Breakin or breakout transmission loss
    Smaller duct dimension, in
    Larger duct dimension, in
    Duct length, ft
    Sheet metal gauge
Calculate:
    TL_{min} = minimum transmission loss
    A_i = duct cross–section area
    A_o = duct sound radiation surface area
    f_L = cutoff frequency for duct breakout
    TL_{out} = duct breakout transmission loss
    $L_{w_r} - L_{w_i}$, dB
    f_1 = cutoff frequency for duct breakin
    $L_{w_t} - L_{w_i}$, dB

COMPUTER PROGRAM – RECTBR.BAS

```
10  ' >RECTANGULAR DUCT BREAKOUT<
20  '
30  DIM T%(8), TT(8), FE(8), MASS(9)
40  PI = 3.141593
50  '
60  '   LIST OF VARIABLES
70  '
80  '   FL = limiting frequency, Hz
90  '   MASS() = mass/unit area of duct walls, lbm/sq. ft.
100 '   DH = large duct dimension, inches
110 '   DL = duct length, feet
120 '   DW = small duct dimension, inches
130 '   FE() = 1/1 octave band center frequency, Hz
140 '   TMAX = maximum transmission loss, dB
150 '   TMIN = minimum transmission loss, dB
160 '   TT() = Transmission loss, dB
170 '   T%() = Sound Power Level, dB
180 '
190 FOR I = 1 TO 9: READ MASS(I): NEXT I
200 '
210 FOR I = 1 TO 8: FE(I) = 62.5 * (2 ^ (I - 1)): NEXT I: ' Frequency - Hz
220 '
230 CLS : PRINT "RECTANGULAR DUCT BREAKOUT AND BREAKIN SOUND TRANS LOSS": PRINT
240 '
250 CLS : PRINT "RECTANGULAR DUCT BREAKOUT AND BREAKIN SOUND TRANSMISSION LOSS"
260 PRINT
270 PRINT "1. DUCT BREAKOUT SOUND TRANSMISSION LOSS"
280 PRINT "2. DUCT BREKIN SOUND TRANSMISSION LOSS"
290 INPUT "SELECTION: ", S1
300 IF S1 >= 1 AND S1 <= 2 THEN GOTO 310 ELSE GOTO 290
310 PRINT
320 INPUT "ENTER SMALLER DUCT DIMENSION (IN): ", DW
330 INPUT "ENTER LARGER DUCT DIMENSION (IN): ", DH
340 IF DW > DH THEN PRINT "REDO": GOTO 320
350 INPUT "ENTER DUCT LENGTH (FT): ", DL
360 '
370 PRINT
380 GOSUB 840: ' Determines sheet metal gauge and mass/unit area
390 PRINT
400 PRINT "CONTINUE? (Y OR N)"
```

```
410 Q$ = INKEY$: IF Q$ = "Y" OR Q$ = "N" OR Q$ = "y" OR Q$ = "n" THEN GOTO 420 ELSE GOTO 410
420 IF Q$ = "N" OR Q$ = "n" THEN GOTO 230
430 '
440 FL = 24134 / SQR(DW * DH): ' Equation 7
450 TMIN = 10 * LOG(24 * DL * ((1 / DW) + (1 / DH))) / LOG(10): ' Equation 9
460 TMAX = 45
470 '
480 FOR I = 1 TO 8
490 '     Equation 6.11
500     IF FE(I) < FL THEN TT(I) = 10 * LOG((FE(I) * MASS(S) ^ 2) / (DW + DH)) / LOG(10) + 17: IF TT(I) < TMIN THEN TT(I) = TMIN
510 '     Equation 10
520     IF FE(I) >= FL THEN TT(I) = 20 * LOG(MASS(S) * FE(I)) / LOG(10) - 31: IF TT(I) > 45 THEN TT(I) = 45
530     T%(I) = TT(I) - 10 * LOG(24 * DL * (DW + DH) / (DW * DH)) / LOG(10): ' Equation 6.4
540 NEXT I
550 '
560 T1$ = "DUCT BREAKOUT LW RED"
570 IF S1 = 1 THEN GOTO 730
580 FL = 6764 / DH: ' Equation 6.14
590 '
600 FOR I = 1 TO 8
610 '     Equation 6.15
620     IF FE(I) <= FL THEN TL1 = TT(I) - 4 - 10 * LOG(DH / DW) / LOG(10) + 20 * LOG(FE(I) / FL) / LOG(10): TL2 = 10 * LOG(12 * DL * ((1 / DW) + (1 / DH))) / LOG(10): GOTO 640
630     GOTO 650
640     IF TL1 >= TL2 THEN TT(I) = TL1 ELSE TT(I) = TL2: GOTO 660
650     IF FE(I) > FL THEN TT(I) = TT(I) - 3: ' Equation 6.16
660     T%(I) = TT(I) + 3: ' Equation 6.6
670 NEXT I
680 '
690 T1$ = "DUCT BREAKIN LW RED"
700 '
710 ' Output statements
720 '
730 PRINT
740 A$ = STRING$(41, CHR$(45)) + " ___ ___ ___ ___ ___ ___ ___ ___"
750 PRINT SPC(45); "1/1 OCT BAND CENTER FREQ"
760 PRINT SPC(10); "DESCRIPTION"; SPC(21); " 63  125  250  500   1K   2K   4K   8K"
770 PRINT A$
780 PRINT USING "\                                        \"; T1$;
790 PRINT USING "  ###"; T%(1); T%(2); T%(3); T%(4); T%(5); T%(6); T%(7); T%(8)
800 END
810 '
820 ' Subroutine for sheet metal gauges and mass/unit area
830 '
840 PRINT "SHEET METAL GAUGES"
850 PRINT "1. 10"; TAB(15); "4. 16"; TAB(30); "7. 22"
860 PRINT "2. 12"; TAB(15); "5. 18"; TAB(30); "8. 24"
870 PRINT "3. 14"; TAB(15); "6. 20"; TAB(30); "9. 26"
880 INPUT "SELECTION: ", S
890 IF S >= 1 AND S <= 9 THEN GOTO 900 ELSE GOTO 880
900 RETURN
910 '
920 DATA 5.625,4.375,3.125,2.5,2.0,1.5,1.25,1.0,0.75: 'Unit densities for sheet metal
```

SAMPLE OUTPUT FOR EXAMPLE 6.1

RECTANGULAR DUCT BREAKOUT AND BREAKIN SOUND TRANSMISSION LOSS

1. DUCT BREAKOUT SOUND TRANSMISSION LOSS
2. DUCT BREAKIN SOUND TRANSMISSION LOSS
SELECTION: 1

ENTER SMALLER DUCT DIMENSION (IN): 12
ENTER LARGER DUCT DIMENSION (IN): 24
ENTER DUCT LENGTH (FT): 20

SHEET METAL GAUGES
1. 10   4. 16   7. 22
2. 12   5. 18   8. 24
3. 14   6. 20   9. 26
SELECTION: 8

CONTINUE? (Y OR N)

DESCRIPTION	63	125	250	500	1K	2K	4K	8K
DUCT BREAKOUT LW RED	2	5	8	11	14	17	23	27

1/1 OCT BAND CENTER FREQ

RECTANGULAR DUCT BREAKOUT AND BREAKIN SOUND TRANSMISSION LOSS

1. DUCT BREAKOUT SOUND TRANSMISSION LOSS
2. DUCT BREAKIN SOUND TRANSMISSION LOSS
SELECTION: 2

ENTER SMALLER DUCT DIMENSION (IN): 12
ENTER LARGER DUCT DIMENSION (IN): 24
ENTER DUCT LENGTH (FT): 20

SHEET METAL GAUGES
1. 10   4. 16   7. 22
2. 12   5. 18   8. 24
3. 14   6. 20   9. 26
SELECTION: 8

CONTINUE? (Y OR N)

DESCRIPTION	63	125	250	500	1K	2K	4K	8K
DUCT BREAKIN LW RED	18	18	20	28	31	35	41	45

1/1 OCT BAND CENTER FREQ

CIRCULAR DUCT BREAKOUT AND BREAKIN

PROGRAM OUTLINE

Read in sheet metal mass/unit area's
Input system parameters:
    Breakout or Breakin transmission loss
    Long seam or spiral wound duct
    Duct diameter, in
    Duct length, ft
    Sheet metal gauge
Calculate:
    A_i = duct cross–section area
    A_o = duct sound radiation surface area
    TL_{out} = duct breakout transmission loss
    $L_{w_r} - L_{w_i}$, dB
    f_1 = cutoff frequency for duct breakin
    $L_{w_t} - L_{w_i}$, dB

COMPUTER PROGRAM – CIRCBR.BAS

```
10  ' >BREAKOUT AND BREAKIN FOR CIRCULAR DUCTS<
20  DIM T%(8), TT(8), FE(8), MASS(9)
30  '
40  '     LIST OF VARIABLES
50  '
60  '     LF = Limiting frequency, Hz
70  '     MASS() = Mass/unit area for sheetmetal, lbm/sq. ft.
80  '     CO = Constant for use with equation 1
90  '     DI = Inside duct diameter, inches
100 '     DL = Duct length, ft
110 '     FE = 1/1 octave band center frequency'
120 '     S1 = Selection for breakout or breakin
130 '     S2 = Selection for long seam or spiral wound ducts
140 '     TL1 = Equation 6.24a
150 '     TL2 = Equation 6.24b
160 '     TT() = Transmission loss, dB
170 '     T%() = Transmission loss output, dB
180 '
190 FOR I = 1 TO 9: READ MASS(I): NEXT I
200 FOR I = 1 TO 7: FE(I) = 62.5 * (2 ^ (I - 1)): NEXT I
210 '
220 ' Subroutine for Circular Duct Breakout and Breakin Sound Transmission Loss
230 '
240 CLS : PRINT "CIRCULAR DUCT BREAKOUT AND BREAKIN SOUND TRANSMISSION LOSS"
250 PRINT
260 PRINT "1. DUCT BREAKOUT SOUND TRANSMISSION LOSS"
270 PRINT "2. DUCT BREAKIN SOUND TRANSMISSION LOSS"
280 INPUT "SELECTION: ", S1
290 IF S1 >= 1 AND S1 <= 2 THEN GOTO 300 ELSE GOTO 280
300 PRINT : PRINT "1. LONG SEAM DUCT    2. SPIRAL WOUND DUCT"
310 INPUT "SELECTION: ", S2
320 IF S2 >= 1 AND S2 <= 2 THEN GOTO 330 ELSE GOTO 310
330 PRINT : INPUT "ENTER DUCT DIAMETER (IN): ", DI
340 INPUT "ENTER DUCT LENGTH (FT): ", DL
350 PRINT
360 GOSUB 850
370 PRINT
380 PRINT "CONTINUE? (Y OR N)"
390 Q$ = INKEY$: IF Q$ = "Y" OR Q$ = "N" OR Q$ = "y" OR Q$ = "n" THEN GOTO 400 ELSE GOTO 390
400 IF Q$ = "N" OR Q$ = "n" THEN GOTO 220
410 '
420 FOR I = 1 TO 7
430    IF S2 = 1 THEN CO = 230.4: ' constant for use with equation 6.19
```

```
440     IF S2 = 2 THEN C0 = 232.9: ' constant for use with equation 6.19
450 '      Equation 6.19
460     TL1 = 17.6 * LOG(MASS(S)) / LOG(10) - 49.8 * LOG(FE(I)) / LOG(10) - 55.3 * LOG(DI) / LOG(10) + C0
470 '      Equation 6.20
480     TL2 = 17.6 * LOG(MASS(S)) / LOG(10) - 6.6 * LOG(FE(I)) / LOG(10) - 36.9 * LOG(DI) / LOG(10) + 97.4
490     IF TL1 > TL2 THEN TT(I) = TL1 ELSE TT(I) = TL2
500 '      Equation 6.21
510     IF DI >= 26 AND FE(I) >= 4000 THEN TT(I) = 17.6 * LOG(MASS(S)) / LOG(10) - 36.9 * LOG(DI) / LOG(10) + 90.6
520     IF TT(I) > 50 THEN TT(I) = 50
530     T%(I) = TT(I) - 10 * LOG(48 * DL / DI) / LOG(10): ' Equation 4
540 NEXT I
550 '
560 T1$ = "DUCT BREAKOUT LW RED"
570 IF S1 = 1 THEN GOTO 750: ' If calculating breakout then skip to print statements
580 '
590 LF = 7929 / DI: ' Equation 6.23
600 '
610 FOR I = 1 TO 7
620 '      Equation 6.24
630     IF FE(I) <= LF THEN TL1 = TT(I) - 4 + 20 * LOG(FE(I) / LF) / LOG(10): TL2 = 10 * LOG(24 * DL / DI) / LOG(10): GOTO 650
640     GOTO 670
650     IF TL1 >= TL2 THEN TT(I) = TL1 ELSE TT(I) = TL2: GOTO 680
660 '      Equation 6.25
670     IF FE(I) > LF THEN TT(I) = TT(I) - 3
680     T%(I) = TT(I) + 3: ' Equation 6.6
690 NEXT I
700 '
710 T1$ = "DUCT BREAKIN LW RED"
720 '
730 ' Output statments
740 '
750 PRINT
760 PRINT SPC(45); "1/1 OCT BAND CENTER FREQ"
770 PRINT SPC(10); "DESCRIPTION"; SPC(21); "  63  125  250  500   1K   2K   4K"
780 PRINT STRING$(41, CHR$(45)) + " ___ ___ ___ ___ ___ ___ ___"
790 PRINT USING "\                                        \"; T1$;
800 PRINT USING "  ###"; T%(1); T%(2); T%(3); T%(4); T%(5); T%(6); T%(7)
810 END
820 '
830 ' Subroutine for sheet metal gauges and mass/unit areas
840 '
850 PRINT "SHEET METAL GAUGES"
860 PRINT "1. 10"; TAB(15); "4. 16"; TAB(30); "7. 22"
870 PRINT "2. 12"; TAB(15); "5. 18"; TAB(30); "8. 24"
880 PRINT "3. 14"; TAB(15); "6. 20"; TAB(30); "9. 26"
890 INPUT "SELECTION: ", S
900 IF S >= 1 AND S <= 9 THEN GOTO 910 ELSE GOTO 890
910 RETURN
920 DATA 5.625,4.375,3.125,2.5,2.0,1.5,1.25,1.0,0.75: ' mass/unit areas
```

SAMPLE OUTPUT FOR EXAMPLE 6.2

CIRCULAR DUCT BREAKOUT AND BREAKIN SOUND TRANSMISSION LOSS

1. DUCT BREAKOUT SOUND TRANSMISSION LOSS
2. DUCT BREAKIN SOUND TRANSMISSION LOSS
SELECTION: 1

1. LONG SEAM DUCT   2. SPIRAL WOUND DUCT
SELECTION: 1

ENTER DUCT DIAMETER (IN): 14
ENTER DUCT LENGTH (FT): 15

SHEET METAL GAUGES
1. 10   4. 16   7. 22
2. 12   5. 18   8. 24
3. 14   6. 20   9. 26
SELECTION: 8

CONTINUE? (Y OR N)

DESCRIPTION	63	125	250	500	1K	2K	4K
DUCT BREAKOUT LW RED	33	33	31	20	18	16	14

(1/1 OCT BAND CENTER FREQ)

CIRCULAR DUCT BREAKOUT AND BREAKIN SOUND TRANSMISSION LOSS

1. DUCT BREAKOUT SOUND TRANSMISSION LOSS
2. DUCT BREAKIN SOUND TRANSMISSION LOSS
SELECTION: 2

1. LONG SEAM DUCT   2. SPIRAL WOUND DUCT
SELECTION: 1

ENTER DUCT DIAMETER (IN): 14
ENTER DUCT LENGTH (FT): 15

SHEET METAL GAUGES
1. 10   4. 16   7. 22
2. 12   5. 18   8. 24
3. 14   6. 20   9. 26
SELECTION: 8

CONTINUE? (Y OR N)

DESCRIPTION	63	125	250	500	1K	2K	4K
DUCT BREAKIN LW RED	30	36	39	35	35	33	31

(1/1 OCT BAND CENTER FREQ)

COMPUTER PROGRAMS FOR CHAPTER 6

FLAT OVAL DUCTS

PROGRAM OUTLINE

Read in sheet metal mass/unit area's for sheet metal
Input system parameters
    Breakin or breakout transmission loss
    Minor axis, inches
    Major axis, inches
    Duct length, ft
Calculate:
Cross sectional area, A_i
    Perimeter, P
    Surface area, A_o
    Fraction of perimeter taken up by flat sides, σ
    Minimum breakout transmission loss, $TL_{out}(min)$
    Breakout transmission loss, TL_{out}
    Limiting frequency, f_L
    Brakin or breakout sound power level, dB
Print out results

COMPUTER PROGRAM – FLOVBR.BAS

```
10  ' >BREAKOUT AND BREAKIN FOR FLAT-OVAL DUCTS COMPUTER PROGRAM<
20  PI = 3.141593
30  DIM TT(8), FE(8), MASS(9), T$(20)
40  '
50  '      LIST OF VARIABLES
60  '
70  '      A = major axis, inches
80  '      AI = cross sectional area, sq. in.
90  '      AO = surface area, sq. in.
100 '      B = minor axis, inches
110 '      LF = limiting frequency, Hz
120 '      MASS() = mass/unit area, lbm/sq. ft.
130 '      DL = duct length, ft
140 '      FE() = 1/1 octave band center frequency, Hz
150 '      P = perimeter, inches
160 '      S1 = selection for breakout or breakin
170 '      SIGMA = fraction of perimeter taken up by the duct sides
180 '      TL1 = See Equation 6.34a
190 '      TL2 = See Equation 6.34b
200 '      T$() = sound power, dB
210 '
220 FOR I = 1 TO 9: READ MASS(I): NEXT I: 'Mass/Unit area for sheet metal
230 FOR I = 1 TO 8: FE(I) = 62.5 * (2 ^ (I - 1)): NEXT I: ' 1/1 octave band center frequency
240 '
250 ' Flat-Oval Duct Breakout and Breakin Sound Transmission Loss
260 '
270 CLS : PRINT "FLAT-OVAL DUCT BREAKOUT AND BREAKIN SOUND TRANSMISSION LOSS"
280 PRINT
290 PRINT "1. DUCT BREAKOUT SOUND TRANSMISSION LOSS"
300 PRINT "2. DUCT BREAKIN SOUND TRANSMISSION LOSS"
310 INPUT "SELECTION: ", S1
320 IF S1 >= 1 AND S1 <= 2 THEN GOTO 330 ELSE GOTO 310
330 PRINT
340 INPUT "MINOR AXIS, (IN): ", B
350 INPUT "MAJOR AXIS, (IN): ", A
360 IF B > A THEN PRINT "MAJOR AXIS MUST BE LARGER THEN MINOR AXIS": GOTO 340
370 INPUT "DUCT LENGTH, (FT): ", DL
380 PRINT
390 GOSUB 890: 'Subroutine for determining sheet metal gage
400 PRINT
410 PRINT "CONTINUE? (Y OR N)"
```

```
420 Q$ = INKEY$: IF Q$ = "Y" OR Q$ = "N" OR Q$ = "y" OR Q$ = "n" THEN GOTO 430 ELSE GOTO 420
430 IF Q$ = "N" OR Q$ = "n" THEN GOTO 270
440 '
450 AO = 12 * DL * (2 * (A - B) + PI * B): ' Equation 6.27
460 AI = B * (A - B) + PI * B ^ 2 / 4: ' Equation 6.26
470 P = 2 * (A - B) + PI * B: ' Equation 6.28
480 SIGMA = 1 / (1 + (PI * B) / (2 * (A - B))): ' Equation 6.29
490 TMIN = 10 * LOG(AO / AI) / LOG(10): ' Equation 6.30
500 LF = 8115 / B: ' Equation 6.32
510 '
520 FOR I = 1 TO 8
530 ' Equation 6.31
540   TT(I) = 10 * LOG((MASS(S) ^ 2 * FE(I)) / (SIGMA ^ 2 * P)) / LOG(10) + 20
550   IF TT(I) < TMIN THEN TT(I) = TMIN
560   IF TT(I) > 50 THEN TT(I) = 50
570   IF FE(I) >= LF THEN T$(I) = " -": GOTO 590
580   T$(I) = STR$(INT(TT(I) - 10 * LOG(AO / AI) / LOG(10))): ' Equation 6.2
590 NEXT I
600 '
610 T1$ = "DUCT BREAKOUT LW RED"
620 IF S1 = 1 THEN GOTO 760
630 BF = 6764 / ((A - B) * SQR(1 + (PI * B) / (2 * (A - B)))): ' Equation 6.33
640 '
650 FOR I = 1 TO 8
660 ' Equation 6.34
670   TL1 = TT(I) + 10 * LOG(FE(I) ^ 2 * AI) / LOG(10) - 81
680   TL2 = 10 * LOG((6 * P * DL) / AI) / LOG(10)
690 ' Equation 6.35
700   IF FE(I) > LF THEN T$(I) = " -": GOTO 740
710   IF FE(I) > BF THEN TT(I) = TT(I) - 3: GOTO 730
720   IF TL1 >= TL2 THEN TT(I) = TL1 ELSE TT(I) = TL2
730   T$(I) = STR$(CINT(TT(I) + 3)): ' Equation 6.6
740 NEXT I
750 T1$ = "DUCT BREAKIN LW RED"
760 '
770 ' Output statements
780 '
790 PRINT
800 PRINT SPC(45); "1/1 OCT BAND CENTER FREQ"
810 PRINT SPC(10); "DESCRIPTION"; SPC(21); " 63 125 250 500  1K  2K  4K  8K"
820 PRINT STRING$(41, CHR$(45)) + " ___ ___ ___ ___ ___ ___ ___ ___"
830 PRINT USING "                                                      \"; T1$;
840 PRINT USING "\    \"; T$(1); T$(2); T$(3); T$(4); T$(5); T$(6); T$(7); T$(8)
850 END
860 '
870 ' Subroutine for sheet metal gauges and mass/unit areas
880 '
890 PRINT "SHEET METAL GAUGES"
900 PRINT "1. 10"; TAB(15); "4. 16"; TAB(30); "7. 22"
910 PRINT "2. 12"; TAB(15); "5. 18"; TAB(30); "8. 24"
920 PRINT "3. 14"; TAB(15); "6. 20"; TAB(30); "9. 26"
930 INPUT "SELECTION: ", S
940 IF S >= 1 AND S <= 9 THEN GOTO 950 ELSE GOTO 930
950 RETURN
960 DATA 5.625,4.375,3.125,2.5,2.0,1.5,1.25,1.0,0.75: ' mass/unit area
```

SAMPLE OUTPUT FOR EXAMPLE 6.3

Flat—Oval Duct Breakout and Breakin Sound Transmission Loss

1. Duct Breakout Sound Transmission Loss
2. Duct Breakin Sound Transmission Loss
SELECTION: 1

MINOR AXIS, (in): 6
MAJOR AXIS, (in): 24
DUCT LENGTH, (ft): 20

Sheet Metal Gauges
1. 10   4. 16   7. 22
2. 12   5. 18   8. 24
3. 14   6. 20   9. 26
SELECTION: 8

CONTINUE? (Y OR N)

DESCRIPTION	1/1 OCT BAND CENTER FREQ						
	63	125	250	500	1K	2K	4K
DUCT BREAKIN LW RED	4	7	10	13	16	—	—

Flat—Oval Duct Breakout and Breakin Sound Transmission Loss

1. Duct Breakout Sound Transmission Loss
2. Duct Breakin Sound Transmission Loss
SELECTION: 2

MINOR AXIS, (in): 6
MAJOR AXIS, (in): 24
DUCT LENGTH, (ft): 20

Sheet Metal Gauges
1. 10   4. 16   7. 22
2. 12   5. 18   8. 24
3. 14   6. 20   9. 26
SELECTION: 8

CONTINUE? (Y OR N)

DESCRIPTION	1/1 OCT BAND CENTER FREQ						
	63	125	250	500	1K	2K	4K
DUCT BREAKOUT LW RED	20	20	22	33	36	—	—

RECTANGLUAR DUCT LAGGING

PROGRAM OUTLINE

Input data for mass/unit areas for duct materials
Input system parameters:
    Duct dimensions
    Thickness of absorbent material
    Sheet metal gauge of duct wall
    Type of outer covering
    Mass/unit area of outer covering
Calculate:
    P_1 = perimeter of the duct
    P_2 = perimeter of the outer covering
    S = cross-sectional area of absorbent material
    IL(lf) = low frequency insertion loss
    f_r = resonance frequency
    IL = Insertion loss

COMPUTER PROGRAM – RECLAG.BAS

```
10  '>INSERTION LOSS OF EXTERNAL LAGGING ON RECTANGULAR DUCTS COMPUTER PROGRAM<
20  '
30  DIM MASS(2, 9), T%(8)
40  '
50  '      LIST OF VARIABLES
60  '
70  '      A = Larger duct dimension, in
80  '      B = 0.71 x resonance frequency, point B on figure 1
90  '      MASS() = Mass/unit area of duct material, lb/sq. ft.
100 '      C = 1.41 x resonance frequency, point C on figure 1
110 '      FR = Resonance frequency, Hz
120 '      FZ = 1/1 octave band center frequency
130 '      H = Thickness of absorbent material
140 '      IL = Insertion loss, dB
150 '      ILLF = Low frequency insertion loss, dB
160 '      LL = Lower limit of 1/1 octave band, Hz
170 '      M1 = Mass/unit area of duct, lb/sq. ft.
180 '      M2 = Mass/unit area of outer covering, lb/sq. ft.
190 '      P1 = Perimeter of uncovered duct, in
200 '      P2 = Perimeter of outer covering, in
210 '      S = Cross-sectional area of absorbent material, sq. in.
220 '      S1 = Selection for rigid or limp outer covering
230 '      S2 = Selection for type of rigid outer covering
240 '      S3 = Selection for sheet metal gauge or Gypsum board
250 '      T%() = 1/1 octave band insertion loss output, dB
260 '      UL = Upper limit of 1/1 octave band, Hz
270 '      W = Smaller duct dimension, in
280 '
290 '
300 FOR I = 1 TO 9
310   READ MASS(1, I): ' Reads in mass/unit areas for sheet metal
320 NEXT I
330 '
340 FOR I = 1 TO 4
350   READ MASS(2, I): ' Reads in mass/unit areas for Gypsum board
360 NEXT I
370 '
380 CLS : PRINT "INSERTION LOSS OF EXTERNAL WALL LAGGING ON RECTANGULAR DUCTS"
390 PRINT
400 INPUT "LARGER DUCT DIMENSION (IN): ", A
410 INPUT "SMALLER DUCT DIMENSION (IN): ", W
420 INPUT "THICKNESS OF SOUND ABSORBING BLANKET (IN): ", H
430 '
440 PRINT
```

COMPUTER PROGRAMS FOR CHAPTER 6

```
450 PRINT "DUCT WALL"
460 GOSUB 1220: ' Subroutine for sheet metal gauges
470 '
480 M1 = MASS(1, S3): ' mass/unit area of duct
490 '
500 PRINT
510 INPUT "TYPE OF OUTER COVERING: (1) RIGID      (2) LIMP : ", S1
520 IF S1 >= 1 AND S1 <= 2 THEN GOTO 530 ELSE GOTO 510
530 PRINT
540 PRINT "TYPE OF MATERIAL FOR EXTERNAL COVERING"
550 '
560 IF S1 = 2 THEN GOTO 680
570 PRINT "1. SHEET METAL"
580 PRINT "2. GYPSUM BOARD"
590 INPUT "SELECTION: ", S2
600 IF S2 >= 1 AND S2 <= 2 THEN GOTO 610 ELSE GOTO 590
610 PRINT
620 PRINT "EXTERNAL COVERING"
630 ON S2 GOSUB 1220, 1320
640 '
650 M2 = MASS(S2, S3)
660 GOTO 700
670 '
680 INPUT "MASS/UNIT AREA OF OUTER COVERING (LB/SQ FT): ", M2
690 '
700 PRINT
710 PRINT "CONTINUE? (Y OR N)"
720 Q$ = INKEY$: IF Q$ = "Y" OR Q$ = "N" OR Q$ = "y" OR Q$ = "n" THEN GOTO 730 ELSE GOTO 720
730 IF Q$ = "N" OR Q$ = "n" THEN GOTO 380
740 PRINT
750 P1 = 2 * (A + W): ' Perimeter of uncovered duct, equation 6.37
760 P2 = 2 * (A + W + 4 * H): ' Perimeter of covered duct, equation 6.38
770 S = 2 * H * (A + W + 2 * H): ' Cross-sectional area of absorbent material, equation 6.40
780 '
790 ILLF = 20 * LOG(1 + (M2 / M1) * (P1 / P2)) / LOG(10): ' Low frequency insertion loss:' eq. 6.36
800 FR = 156 * SQR((P2 / P1 + M2 / M1) * P1 / (M2 * S)): ' Resonance frequency:' Equation 6.39
810 B = .71 * FR:   ' Point B on figure 6.6a
820 C = 1.41 * FR: ' Point C on figure 6.6a
830 '
840 IF S1 = 2 THEN GOTO 1030: ' S1 = 2 when outer covering is limp
850 '
860 ' Rigid Outer Covering
870 '
880 FOR I = 1 TO 8
890    FZ = 62.5 * (2 ^ (I - 1)): ' 1/1 octave band center frequency
900    LL = 44 * 2 ^ (I - 1): UL = 88 * 2 ^ (I - 1)
910    IF FR > LL AND FR < UL THEN IL = ILLF - 5: GOTO 970
920    IF FZ <= B THEN IL = ILLF
930    IF FZ > B AND FZ <= FR THEN IL = ILLF - 67.231 * LOG(FZ / B) / LOG(10): ' Eq. 6.41
940    IF FZ > FR AND FZ <= C THEN IL = ILLF - 10 + 67.016 * LOG(FZ / FR) / LOG(10): ' Eq. 6.42
950    IF FZ > C THEN IL = ILLF + 29.897 * LOG(FZ / C) / LOG(10): ' Equation 6.6.43
960    IF IL > 25 THEN IL = 25
970    T%(I) = IL
980 NEXT I
990 GOTO 1130
1000 '
1010 ' Limp Outer covering
1020 '
1030 FOR I = 1 TO 8
1040    FZ = 62.5 * (2 ^ (I - 1)): ' 1/1 octave band center frequency
1050    IF FZ <= FR THEN IL = ILLF
1060    IF FZ > FR THEN IL = 29.897 * LOG(FZ / FR) / LOG(10) + ILLF: ' Equation 6.44
1070    IF IL > 25 THEN IL = 25
1080    T%(I) = IL
1090 NEXT I
1100 '
1110 ' Output statements
1120 '
1130 PRINT SPC(45); "1/1 OCT BAND CENTER FREQ"
1140 PRINT SPC(10); "DESCRIPTION"; SPC(21); " 63 125 250 500  1K  2K  4K  8K"
1150 PRINT STRING$(41, CHR$(45)) + " ___ ___ ___ ___ ___ ___ ___ ___"
1160 PRINT USING "\                                        \"; "Insertion Loss, dB";
1170 PRINT USING "  ###"; T%(1); T%(2); T%(3); T%(4); T%(5); T%(6); T%(7); T%(8)
1180 END
```

```
1190 '
1200 ' Subroutine for sheet metal gauges and corresponding mass/unit area
1210 '
1220 PRINT "SHEET METAL GAUGES"
1230 PRINT "1. 10"; TAB(15); "4. 16"; TAB(30); "7. 22"
1240 PRINT "2. 12"; TAB(15); "5. 18"; TAB(30); "8. 24"
1250 PRINT "3. 14"; TAB(15); "6. 20"; TAB(30); "9. 26"
1260 INPUT "SELECTION: ", S3
1270 IF S3 >= 1 AND S3 <= 9 THEN GOTO 1280 ELSE GOTO 1260
1280 RETURN
1290 '
1300 ' Subroutine for Gypsum board thickness and corresponding mass/unit area
1310 '
1320 PRINT "GYPSUM BOARD"
1330 PRINT "1. 1 LAYER 3/8 IN"
1340 PRINT "2. 1 LAYER 1/2 IN"
1350 PRINT "3. 1 LAYER 5/8 IN"
1360 PRINT "4. 2 LAYERS 1/2 IN"
1370 INPUT "SELECTION: ", S3
1380 IF S3 >= 1 AND S3 <= 4 THEN GOTO 1390 ELSE GOTO 1370
1390 RETURN
1400 '
1410 DATA 5.625,4.375,3.125,2.5,2.0,1.5,1.25,1.0,0.75: ' Sheet Metal mass/unit area, lb/sq. ft.
1420 DATA 1.6,2.1,2.7,4.6: ' Gypsum Board mass/unit areas, lb/sq. ft.
```

SAMPLE OUTPUT FOR EXAMPLE 6.4

INSERTION LOSS OF EXTERNAL WALL LAGGING ON RECTANGULAR DUCTS

LARGER DUCT DIMENSION (IN): 8
SMALLER DUCT DIMENSION (IN): 8
THICKNESS OF SOUND ABSORBING BLANKET (IN): 1

DUCT WALL
SHEET METAL GAUGES
1. 10   4. 16   7. 22
2. 12   5. 18   8. 24
3. 14   6. 20   9. 26
SELECTION: 5

TYPE OF OUTER COVERING: (1) RIGID   (2) LIMP : 1

TYPE OF MATERIAL FOR EXTERNAL COVERING
1. SHEET METAL
2. GYPSUM BOARD
SELECTION: 2

EXTERNAL COVERING
GYPSUM BOARD
1. 1 LAYER 3/8 IN
2. 1 LAYER 1/2 IN
3. 1 LAYER 5/8 IN
4. 2 LAYERS 1/2 IN
SELECTION: 2

CONTINUE? (Y OR N)

DESCRIPTION	63	125	250	500	1K	2K	4K	8K
Insertion Loss, dB	5	0	7	16	25	25	25	25

1/1 OCT BAND CENTER FREQ

COMPUTER PROGRAMS FOR CHAPTER 7

APPENDIX 6

COMPUTER PROGRAMS FOR CHAPTER 7

A6–1

TRANSMISSION LOSS THROUGH CEILING SYSTEMS

PROGRAM OUTLINE

Read in insertion loss data
Input System Parameters:
    Type of ceiling system
    Type of suspended ceiling
Calculate:
    Transmission loss through ceiling

COMPUTER PROGRAM – CEILING.BAS

```
10  ' >CEILING SYSTEM TRANSMISSION LOSS COMPUTER PROGRAM<
20  DIM T%(7), TT(7), D(10, 7), TL(7)
30  '
40  '     LIST OF VARIABLES
50  '
60  '     TL() = Transmission Loss, dB
70  '     D()  = Insertion Loss Values, dB
80  '     RAT  = Insertion Loss Reduction Coefficients for flanking and acoustic leaks
90  '     S1   = Selection for type of ceiling system
100 '     S2   = Selection for type of suspended ceiling
110 '     T%() = Transmission loss output, dB
120 '
130 FOR I = 1 TO 10
140   FOR J = 1 TO 7: ' Data only available to 4000 Hz
150     READ D(I, J): ' Reads in insertion loss values
160   NEXT J
170 NEXT I
180 '
190 ' Menu for type ceiling system
200 '
210 CLS : PRINT "TRANSMISSION LOSS ASSOCIATED WITH CEILING SYSTEMS": PRINT
220 PRINT "TL VALUES OF CEILING SYSTEMS": PRINT
230 PRINT " 1. 1 LAYER 3/8 IN GYPSUM BOARD; SURFACE WEIGHT = 1.6 LB/SQ FT"
240 PRINT " 2. 1 LAYER 1/2 IN GYPSUM BOARD; SURFACE WEIGHT = 2.1 LB/SQ FT"
250 PRINT " 3. 1 LAYER 5/8 IN GYPSUM BOARD; SURFACE WEIGHT = 2.7 LB/SQ FT"
260 PRINT " 4. 2 LAYERS 1/2 IN GYPSUM BOARD; SURFACE WEIGHT = 4.6 LB/SQ FT"
270 PRINT
280 PRINT "ACOUSTICAL CEILING TILE – EXPOSED T-BAR GRID SUSPENDED LAY-IN CEILINGS"
290 PRINT " 5. 2 FT X 4 FT X 5/8 IN LAY-IN TILE; SURFACE WEIGHT = 0.6-0.7 LB/SQ FT"
300 PRINT " 6. 2 FT X 4 FT X 1-1/2 IN LAY-IN TILE; SURFACE WEIGHT = 0.9 LB/SQ FT"
310 PRINT " 7. 2 FT X 4 FT X 5/8 IN LAY-IN TILE; SURFACE WEIGHT = 0.95-1.1 LB/SQ FT"
320 PRINT " 8. 2 FT X 2 FT X 5/8 IN LAY-IN TILE; SURFACE WEIGHT = 1.2-1.3 LB/SQ FT"
330 PRINT
340 PRINT "ACOUSTICAL CEILING TILE – CONCEALED SPLINE SUSPENDED CEILING"
350 PRINT " 9. 1 FT X 1 FT X 5/8 IN TILE; SURFACE WEIGHT = 1.2 LB/SQ FT"
360 PRINT
370 PRINT "10. ASHRAE 1987 GENERAL CEILING TILE INSERTION LOSS VALUES"
380 PRINT
390 INPUT "SELECTION: ", S1
400 IF S1 >= 1 AND S1 <= 10 THEN GOTO 410 ELSE GOTO 390
410 '
420 ' Menu for type of suspended ceiling
430 '
440 PRINT
450 PRINT "TYPE OF SUSPENDED CEILING": PRINT
460 PRINT "1. GYPBOARD – NO CEILING DIFFUSERS OR PENETRATIONS"
470 PRINT "2. GYPBOARD – FEW CEILING DIFFUSERS AND PENETRATIONS WELL SEALED"
480 PRINT "3. LAY-IN SUSPENDED TILE – NO INTEGRATED LIGHTING OR DIFFUSER SYSTEM"
490 PRINT "4. LAY-IN SUSPENDED TILE – INTEGRATED LIGHTING AND DIFFUSER SYSTEM"
500 PRINT
510 INPUT "SELECTION: ", S2
520 IF S2 >= 1 AND S2 <= 4 THEN GOTO 530 ELSE GOTO 510
530 PRINT
540 PRINT "CONTINUE? (Y OR N)"
550 Q$ = INKEY$: IF Q$ = "Y" OR Q$ = "N" OR Q$ = "y" OR Q$ = "n" THEN GOTO 560 ELSE GOTO 550
560 IF Q$ = "N" OR Q$ = "n" THEN GOTO 210
```

```
570 '
580 FOR I = 1 TO 7: TL(I) = D(S1, I): NEXT I
590 '
600 GOSUB 750
610 FOR I = 1 TO 7: T%(I) = TL(I): NEXT I
620 '
630 ' Output statements
640 '
650 PRINT
660 PRINT SPC(45); "1/1 OCT BAND CENTER FREQ"
670 PRINT SPC(10); "DESCRIPTION"; SPC(21); " 63  125  250  500   1K   2K   4K"
680 PRINT STRING$(41, CHR$(45)) + " ___  ___  ___  ___  ___  ___  ___"
690 PRINT USING "\                                                \"; "TRANSMISSION LOSS OF CEILING SYSTEM, dB";
700 PRINT ; USING " ###"; T%(1); T%(2); T%(3); T%(4); T%(5); T%(6); T%(7)
710 END
720 '
730 ' Transmission Loss Reduction Coefficients for flanking and acoustic leaks
740 '
750 IF S2 = 1 THEN RAT = .0001
760 IF S2 = 2 THEN RAT = .001
770 IF S2 = 3 THEN RAT = .001
780 IF S2 = 4 THEN RAT = .03
790 '
800 FOR I = 1 TO 7
810 TL(I) = -10 * LOG((1 - RAT) * 10 ^ (-TL(I) / 10) + RAT) / LOG(10): ' Equation 7.1
820 NEXT I
830 RETURN
840 '
850 ' Transmission Loss Data, dB
860 '
870 DATA 6,11,17,22,28,32,24: ' 1 LAYER 3/8 IN GYPSUM BOARD; SURFACE WEIGHT = 1.6 LB/SQ FT"
880 DATA 9,14,20,24,30,31,27: ' 1 LAYER 1/2 IN GYPSUM BOARD; SURFACE WEIGHT = 2.1 LB/SQ FT"
890 DATA 10,15,22,26,31,28,30: ' 1 LAYER 5/8 IN GYPSUM BOARD; SURFACE WEIGHT = 2.7 LB/SQ FT"
900 DATA 13,18,26,30,30,29,37: ' 2 LAYERS 1/2 IN GYPSUM BOARD; SURFACE WEIGHT = 4.6 LB/SQ FT"
910 DATA 4,9,9,14,19,24,26: ' 2 FT X 4 FT X 5/8 IN LAY-IN TILE; SURFACE WEIGHT = 0.6-0.7 LB/SQ FT"
920 DATA 4,9,10,11,15,20,25: ' 2 FT X 4 FT X 1-1/2 IN LAY-IN TILE; SURFACE WEIGHT = 0.9 LB/SQ FT"
930 DATA 5,11,13,15,21,24,28: ' 2 FT X 4 FT X 5/8 IN LAY-IN TILE; SURFACE WEIGHT = 0.95-1.1 LB/SQ FT"
940 DATA 5,10,11,15,19,22,24: ' 2 FT X 2 FT X 5/8 IN LAY-IN TILE; SURFACE WEIGHT = 1.2-1.3 LB/SQ FT"
950 DATA 6,14,14,18,22,27,30: ' 1 FT X 1 FT X 5/8 IN TILE; SURFACE WEIGHT = 1.2 LB/SQ FT"
960 DATA 1,2,4,8,9,9,14: ' ASHRAE 1987 GENERAL CEILING TILE INSERTION LOSS VALUES"
```

SAMPLE OUTPUT FOR EXAMPLE 7.1

TRANSMISSION LOSS ASSOCIATED WITH CEILING SYSTEMS

TL VALUES OF CEILING SYSTEMS

1. 1 LAYER 3/8 IN GYPSUM BOARD; SURFACE WEIGHT = 1.6 LB/SQ FT
2. 1 LAYER 1/2 IN GYPSUM BOARD; SURFACE WEIGHT = 2.1 LB/SQ FT
3. 1 LAYER 5/8 IN GYPSUM BOARD; SURFACE WEIGHT = 2.7 LB/SQ FT
4. 2 LAYERS 1/2 IN GYPSUM BOARD; SURFACE WEIGHT = 4.6 LB/SQ FT

ACOUSTICAL CEILING TILE — EXPOSED T-BAR GRID SUSPENDED LAY-IN CEILINGS
5. 2 FT X 4 FT X 5/8 IN LAY-IN TILE; SURFACE WEIGHT = 0.6-0.7 LB/SQ FT
6. 2 FT X 4 FT X 1-1/2 IN LAY-IN TILE; SURFACE WEIGHT = 0.9 LB/SQ FT
7. 2 FT X 4 FT X 5/8 IN LAY-IN TILE; SURFACE WEIGHT = 0.95-1.1 LB/SQ FT
8. 2 FT X 2 FT X 5/8 IN LAY-IN TILE; SURFACE WEIGHT = 1.2-1.3 LB/SQ FT

ACOUSTICAL CEILING TILE — CONCEALED SPLINE SUSPENDED CEILING
9. 1 FT X 1 FT X 5/8 IN TILE; SURFACE WEIGHT = 1.2 LB/SQ FT

10. ASHRAE 1987 GENERAL CEILING TILE INSERTION LOSS VALUES

SELECTION: 2

TYPE OF SUSPENDED CEILING

1. GYPBOARD — NO CEILING DIFFUSERS OR PENETRATIONS
2. GYPBOARD — FEW CEILING DIFFUSERS AND PENETRATIONS WELL SEALED
3. LAY-IN SUSPENDED TILE — NO INTEGRATED LIGHTING OR DIFFUSER SYSTEM
4. LAY-IN SUSPENDED TILE — INTEGRATED LIGHTING AND DIFFUSER SYSTEM

SELECTION: 2

CONTINUE? (Y OR N)

DESCRIPTION	63	125	250	500	1K	2K	4K
Transmission Loss, dB	9	14	20	23	27	28	25

1/1 OCT BAND CENTER FREQ

… A6—4 APPENDIX 6

RECEIVER ROOM SOUND CORRECTIONS

PROGRAM OUTLINE

Read in Data for absorption coefficients
Input system parameters:
    Type of correction method
    receiving room height, ft
    receiving room length, ft
    receiving room width, ft
    location in room regarding directivity
    acoustic characteristics of room
Calculate:
    1/1 octave band receiver room sound corrections
Print out results

COMPUTER PROGRAM – RECCOR.BAS

```
10  ' >Receiving Room Sound Correction<
20  DIM T$(60), T%(7), TT(7), D(42, 7), FE(7), R(7), BP(3)
30  '       LIST OF VARIABLES
40  '
50  '       AR = surface area of room, sq. ft.
60  '       BP() = Directivity Factors
70  '       CH = Ceiling Height, ft
80  '       D() = array used for reading in absorption coefficients
90  '       DS = Average Distance between sound source and receiver, ft
100 '       FE = 1/1 Octave Band Center Frequency
110 '       ND = Number of Sound Sources
120 '       Q = Directiviy Factor
130 '       R() = air absorption coefficients
140 '       RH = Receiving Room Height, ft
150 '       RL = Receiving Room Length, ft
160 '       RR = Average Distance Between Sound Source And Receiver, ft
170 '       RW = Receiving Room Width, ft
180 '       SL = Length of Sound Sources
190 '       S1 = Correction Method to be Used
200 '       S2 = Location of Sound Source(s)
210 '       S3 = Selection for acoustic characteristics of the room
220 '       S4 = Selection for individual sound sources or ceiling diffusers
230 '       TT() = temporary array
240 '       T%() = Lp - Lw, dB
250 '       T1$ = character string used for output
260 '       VOL = Volume of Room, cu. ft.
270 '
280 PI = 3.141593
290 '
300 FOR I = 1 TO 5
310   FOR J = 1 TO 7
320     READ D(I, J)
330   NEXT J
340 NEXT I
350 '
360 FOR I = 1 TO 3: READ BP(I): NEXT I
370 '
380 FE(1) = 63: FE(2) = 125: FE(3) = 250: FE(4) = 500: FE(5) = 1000: FE(6) = 2000: FE(7) = 4000
390 '
400 CLS : LOCATE 2, 20: PRINT "RECEIVER ROOM SOUND CORRECTION MENU": PRINT
410 PRINT "1. MODIFIED DIFFUSE FIELD THEORY ROOM EQUATION - POINT SOURCE (BY THOMPSON)"
420 PRINT "     FOR LARGE IRREGULARLY SHAPED ROOMS THAT DO NOT HAVE EVENLY DISTRIBUTED"
430 PRINT "     SOUND ABSORPTION OR A ROOM DIMENSION RATIO OF 1:1.5:2": PRINT
440 PRINT "2. 1987 ASHRAE ROOM EQUATION - POINT SOURCE (BY SHULTZ)"
450 PRINT "     FOR ORDINARY FURNISHED DWELLINGS AND CONTAINED OFFICES": PRINT
460 PRINT "3. DIFFUSE FIELD THEORY ROOM EQUATION - LINE SOURCE"
470 PRINT "     FOR ROOMS WITH EVENLY DISTRIBUTED SOUND ABSORPTION AND ROOM DIMENSION"
480 PRINT "     RATIO OF 1:1.5:2": PRINT
```

COMPUTER PROGRAMS FOR CHAPTER 7

```
490 INPUT "SELECTION: ", S1
500 '
510 GOSUB 1290
520 '
530 PRINT : INPUT "ENTER RECEIVING ROOM HEIGHT (FT): ", RH
540 INPUT "ENTER RECEIVING ROOM LENGTH (FT): ", RL
550 INPUT "ENTER RECEIVING ROOM WIDTH (FT): ", RW
560 IF S1 = 1 OR S1 = 3 THEN GOSUB 650
570 IF S1 = 2 THEN GOSUB 1040
580 PRINT
590 PRINT SPC(54); "1/1 OCT BAND CENTER FREQ"
600 PRINT SPC(15); "DESCRIPTION"; SPC(26); "  63  125  250  500   1K   2K   4K"
610 PRINT STRING$(51, CHR$(45)) + " ___  ___  ___  ___  ___  ___  ___"
620 PRINT USING "\                                                 \"; T1$;
630 PRINT SPC(3); USING " ###"; T%(1); T%(2); T%(3); T%(4); T%(5); T%(6); T%(7)
640 END
650 PRINT : INPUT "ENTER AVERAGE DIST BETWEEN SOUND SOURCE AND RECEIVER (FT): ", DS
660 IF S1 = 1 THEN INPUT "ENTER NUMBER OF SOUND SOURCES: ", ND
670 IF S1 = 3 THEN INPUT "ENTER LENGTH OF LINE SOURCE (FT): ", SL
680 PRINT : PRINT "LOCATION OF SOUND SOURCE(S)":
690 PRINT "1. ON EXTENDED FLAT SURFACE"
700 PRINT "2. NEAR THE INTERSECTION OF TWO FLAT SURFACES"
710 PRINT "3. NEAR THE INTERSECTION OF THREE FLAT SURFACES"
720 INPUT "SELECTION: ", S2
730 '
740 LOCATE 16, 22: PRINT "ACOUSTIC CHARACTERISTICS OF THE ROOM"
750 PRINT "1. DEAD:         Acoustical ceiling, plush carpet, soft furnishings, people"
760 PRINT "2. MEDIUM DEAD:  Acoustical ceiling, commercial carpet, people"
770 PRINT "3. AVERAGE:      Acoustical ceiling or commercial carpet, people"
780 PRINT "4. MEDIUM LIVE:  Some acoustical materials, people"
790 PRINT "5. LIVE:         People only"
800 INPUT "SELECTION: ", S3
810 PRINT
820 PRINT "CONTINUE? (Y OR N)"
830 Q$ = INKEY$: IF Q$ = "Y" OR Q$ = "N" OR Q$ = "y" OR Q$ = "n" THEN GOTO 840 ELSE GOTO 830
840 IF Q$ = "N" OR Q$ = "n" THEN GOTO 400
850 '
860 Q = BP(S2): AR = 2 * (RL * RW + RL * RH + RW * RH): VOL = RH * RL * RW
870 '
880 FOR I = 1 TO 5: R(I) = 0: NEXT I: R(6) = .0009: R(7) = .0029
890 '
900 FOR I = 1 TO 7
910 AB = D(S3, I) + 4 * R(I) * VOL / AR: RC = AR * AB / (1 - AB): MFP = 4 * VOL / (AR * DS)
920 '
930 IF S1 = 3 THEN TT(I) = 10 * LOG(4 / RC + Q / (2 * PI * DS * SL)) / LOG(10) + 10.5
940 IF S1 = 1 THEN TT(I) = 10 * LOG(MFP * 4 / (DS * AR * AB) + Q * EXP(-R(I) * DS) / (4 * PI * DS ^ 2)) / LOG(10) + 10.5 + 10 * LOG(ND) / LOG(10)
950 T%(I) = TT(I)
960 NEXT I
970 '
980 IF ND = 1 THEN SOU$ = " SOURCE"
990 IF ND > 1 THEN SOU$ = " SOURCES"
1000 ND$ = STR$(ND)
1010 IF S1 = 1 THEN T1$ = "MOD DIFFUSE ROOM CORR -" + ND$ + SOU$
1020 IF S1 = 3 THEN T1$ = "DIFFUSE ROOM CORR - LINE SOURCE"
1030 RETURN
1040 PRINT : PRINT "1. INDIVIDUAL SOUND SOURCES"; SPC(10); "2. DISTRIBUTED CEILING DIFFUSERS"
1050 INPUT "SELECTION: ", S4
1060 IF S4 = 1 THEN CH = 0: ND = 0: AR = 0: GOSUB 1220
1070 IF S4 = 2 THEN RR = 0: ND = 0: GOSUB 1250
1080 PRINT
1090 PRINT "CONTINUE? (Y OR N)"
1100 Q$ = INKEY$: IF Q$ = "Y" OR Q$ = "N" OR Q$ = "y" OR Q$ = "n" THEN GOTO 1110 ELSE GOTO 1100
1110 IF Q$ = "N" OR Q$ = "n" THEN GOTO 400
1120 IF S4 = 2 THEN AR = AR / CH ^ 2
1130 VOL = RH * RL * RW
1140 ND$ = STR$(ND)
1150 FOR I = 1 TO 7
1160 IF S4 = 1 THEN T%(I) = 25 - 5 * LOG(VOL) / LOG(10) - 3 * LOG(FE(I)) / LOG(10) - 10 * LOG(RR) / LOG(10) + 10 * LOG(ND) / LOG(10)
1170 IF S4 = 2 THEN T%(I) = 30 - 5 * LOG(AR) / LOG(10) - 27.6 * LOG(CH) / LOG(10) + 1.3 * LOG(ND) / LOG(10) - 3 * LOG(FE(I)) / LOG(10)
1180 NEXT I
1190 IF S4 = 1 THEN T1$ = "ASHRAE ROOM CORR - " + ND$ + " IND SOUND SOURCES"
1200 IF S4 = 2 THEN T1$ = "ASHRAE ROOM CORR - " + ND$ + " CEILING DIFF'R ARRAY"
```

```
1210 RETURN
1220 PRINT : INPUT "ENTER AVERAGE DISTANCE BETWEEN SOUND SOURCE AND RECEIVER (FT): ", RR
1230 INPUT "ENTER NUMBER OF INDIVIDUAL SOUND SOURCES: ", ND
1240 RETURN
1250 INPUT "ENTER CEILING HEIGHT (FT): ", CH
1260 INPUT "ENTER NUMBER OF CEILING DIFFUSERS: ", ND
1270 INPUT "ENTER AREA SERVED BY EACH DIFF'R (SQ'FT): ", AR
1280 RETURN
1290 CLS : IF S1 = 3 THEN PRINT "DIFFUSE FIELD THEORY ROOM EQUATION - LINE SOURCE"
1300 IF S1 = 1 THEN PRINT "MODIFIED DIFFUSE FIELD THEORY ROOM EQUATION - POINT SOURCE"
1310 IF S1 = 2 THEN PRINT "1987 ASHRAE ROOM EQUATION - POINT SOURCE"
1320 RETURN
1330 '
1340 DATA .26,.30,.35,.4,.43,.46,.52: 'Dead Room
1350 DATA .24,.22,.18,.25,.3,.36,.42: 'Med. Dead Room
1360 DATA .25,.23,.17,.2,.24,.29,.34: 'Average
1370 DATA .25,.23,.15,.15,.17,.2,.23: 'Med. Live Room
1380 DATA .26,.24,.12,.1,0.09,.11,.13: ' Live Room
1390 DATA 2,4,8
```

SAMPLE OUTPUT FOR EXAMPLE 7.2

PRINT "RECEIVER ROOM SOUND CORRECTION MENU

1. MODIFIED DIFFUSE FIELD THEORY ROOM EQUATION - POINT SOURCE (BY THOMPSON)
   FOR LARGE IRREGULARLY SHAPED ROOMS THAT DO NOT HAVE EVENLY DISTRIBUTED
   SOUND ABSORPTION OR A ROOM DIMENSION RATIO OF 1:1.5:2

2. 1987 ASHRAE ROOM EQUATION - POINT SOURCE (BY SHULTZ)
   FOR ORDINARY FURNISHED DWELLINGS AND CONTAINED OFFICES

3. DIFFUSE FIELD THEORY ROOM EQUATION - LINE SOURCE
   FOR ROOMS WITH EVENLY DISTRIBUTED SOUND ABSORPTION AND ROOM DIMENSION
   RATIO OF 1:1.5:2

SELECTION: 2

1987 ASHRAE ROOM EQUATION - POINT SOURCE

ENTER RECEIVER ROOM HEIGHT (FT): 8
ENTER RECEIVER ROOM LENGTH (FT): 15
ENTER RECEIVER ROOM WIDTH (FT): 10

1. INDIVIDUAL SOUND SOURCES   2. DISTRIBUTED CEILING DIFFUSERS
SELECTION: 1

ENTER AVERAGE DISTANCE BETWEEN SOUND SOURCE AND RECEIVER (FT): 8
ENTER NUMBER OF INDIVIDUAL SOUND SOURCES: 1

CONTINUE? (Y OR N)

| | 1/1 OCT BAND CENTER FREQ |||||||
	63	125	250	500	1K	2K	4K
ASHRAE ROOM CORR - 1 IND SOUND SOURCES	−5	−6	−7	−8	−8	−9	−10

SOUND TRANSMISSION THROUGH MECHANICAL EQUIPMENT ROOM WALLS, FLOOR, OR CEILING

PROGRAM OUTLINE

Input Data from Tables 7.5 through 7.9
Input system parameters:
    Mechanical room dimensions
    Distance between sound source and wall
    Sound absorption associated with construction of walls
    Type and amount of acoustical absorbing material added
    Dimensions of common wall
    Type of wall
    Construction quality of wall
    Dimensions of wall
    Adjacent room dimensions
    Distance between receiving point and wall
    Acoustical characteristics of adjacent room
Calculate:
    Surface areas of mechanical room
    Average absorption coefficient of mechanical room
    Change in sound power level from sound source to common wall, dB
    Transmission loss through common wall
    Surface areas in adjacent room
    Change in sound power level thourgh wall to sound pressure level in adjacent room
    Change in sounce sound power in mechanical room to sound pressure level in adjacent room

COMPUTER PROGRAM – MECHRM.BAS

```
10  ' >TRANS LOSS THROUGH MECHANICAL ROOM WALL/FLOOR/CEILING COMPUTER PROGRAM<
20  DIM T1%(7), T2%(7), T3%(7), D(36, 7), FE(7), R(7), AIR(7), TL(31)
30  PI = 3.141593
40  '
50  '    LIST OF VARIABLES
60  '
70  '    A = Area of the surface of hemisphere of radius r, sq ft
80  '    AB = Average absorption coefficient for receiving room
90  '    ACM = Surface area of ceiling and floor of mechanical room, sq ft
100 '    ARM = Surface area of mechanical room, sq ft
110 '    ARR = Surface area of receiver room, sq ft
120 '    ATTN = For use with receiver room sound power level correction
130 '    ATTNW = For use with receiver room sound power level correction
140 '    AVM = Surface area of mechancal room walls, sq ft
150 '    TL() = Transmission Loss Value, dB
160 '    D() = Array used to store absorption coefficients and transmission loss values
170 '    AIR() = Absorption coefficient of air
180 '    FE() = 1/1 Octave Band Center Frequency
190 '    LD = Larger dimension of common wall, ft
200 '    MFP = Mean free path
210 '    PCM = Percentage of surface area covered by acoustical material
220 '    RAT = Correction factor for acoustic leaks and flanking transmission
230 '    R() = average absorption coefficient
240 '    RC = Room constant
250 '    RHM = Mechanical room height, ft
260 '    RHR = Receiver room height, ft
270 '    RLM = Mechanical room length, ft
280 '    RLR = Receiver room length, ft
290 '    RWM = Mechanical room width, ft
300 '    RWR = Recever room width, ft
310 '    RRM = Distance between sound source and floor or wall
320 '    RRR = Distance from wall or ceiling to receiver, ft
```

```
330 '      S1 = Selection for wall and floor configurations
340 '      S2 = Selection for wall construction
350 '      S3 = Selection for acoustical absorbing material (yes or no)
360 '      S4 = Selection for type of acoustical absorbing material
370 '      S6 = Selection for acoustic characteristics of receiver room
380 '      S7 = Selection for dry wall configuration
390 '      S8 = Selection for Masonry Wall/Floor/Ceiling Configurations
400 '      S9 = Selection for quality of construction of mechanical room wall/floor/ceiling
410 '      S10 = Selection for Masonry Wall with resiliently mounted gypboard Configurations
420 '      SD = Smaller dimension of common wall, ft
430 '      T%() = Logarithmic average of sound power reductions
440 '      T1%() = Change in sound power level from source to wall
450 '      T2%() = Transmission loss through mechanical room wall
460 '      T3%() = Sound power level to sound pressure level in receiver room
470 '      TT = Temporary value for use in sound power level calculations
480 '      VOLR = Volume of receiver room, cu ft
490 '      WA = Surface area of mechanical room wall adjacent to receiving room, sq ft
500 '
510 FOR I = 1 TO 36
520   FOR J = 1 TO 7
530     READ D(I, J)
540   NEXT J
550 NEXT I
560 '
570 FOR I = 1 TO 7: FE(I) = 62.5 * 2 ^ (I - 1): NEXT I
580 '
590 CLS
600 PRINT "SOUND TRANS THROUGH MECHANICAL EQUIPMENT ROOM FLOOR/CEILING/WALLS"
610 PRINT
620 PRINT "MECHANICAL EQUIPMENT ROOM DIMENSIONS"
630 INPUT "ENTER MECHANICAL ROOM HEIGHT (FT): ", RHM
640 INPUT "ENTER MECHANICAL ROOM LENGTH (FT): ", RLM
650 INPUT "ENTER MECHANICAL ROOM WIDTH (FT): ", RWM
660 INPUT "ENTER DISTANCE BETWEEN SOUND SOURCE AND FLOOR OR WALL OF MECH RM (FT): ", RRM
670 PRINT
680 : PRINT "SOUND ABSORPTION ASSOCIATED WITH CONSTRUCTION OF WALLS"
690 PRINT "1. POURED CONCRETE"
700 PRINT "2. CONCRETE BLOCK:   PAINTED"
710 PRINT "3. CONCRETE BLOCK:   UNPAINTED"
720 PRINT "4. SINGLE LAYER 5/8 IN GYPSUM BOARD"
730 PRINT "5. DOUBLE LAYER 5/8 IN GYPSUM BOARD"
740 INPUT "SELECTION: ", S2
750 IF S2 >= 1 AND S2 <= 5 THEN GOTO 760 ELSE GOTO 740
760 PRINT
770 PRINT "ADD ACOUSTICAL ABSORBING MATERIAL TO WALLS AND CEILING?   (1. YES    2. NO)"
780 INPUT "SELECTION: ", S3
790 IF S3 >= 1 AND S3 <= 2 THEN GOTO 800 ELSE GOTO 780
800 IF S3 = 2 THEN S4 = 0: PCM = 0: GOTO 890
810 PRINT
820 PRINT "1. 2 IN 3 LB/CU FT DENSITY FIBERGLASS INSULATION"
830 PRINT "2. 3 IN 3 LB/CU FT DENSITY FIBERGLASS INSULATION"
840 PRINT "3. 4 IN 3 LB/CU FT DENSITY FIBERGLASS INSULATION"
850 INPUT "SELECTION: ", S4
860 IF S3 >= 1 AND S3 <= 3 THEN GOTO 870 ELSE GOTO 850
870 PRINT
880 INPUT "ENTER PERCENT WALLS, FLOOR AND CEILING COVERED BY ACOUSTICAL MATERIAL: ", PCM
890 PRINT
900 PRINT "TRANSMISSION LOSS VALUES OF DIFFERENT WALL AND FLOOR CONFIGURATIONS"
910 PRINT "1. TL VALUES OF DRY-WALL WALL CONFIGURATIONS"
920 PRINT "2. TL VALUES OF MASONRY WALL/FLOOR/CEILING CONFIGURATIONS"
930 PRINT "3. TL VALUES OF PAINTED MASONRY BLOCK WALLS AND PAINTED"
940 PRINT "   BLOCK WALLS WITH RESILIENTLY MOUNTED GYPSUM WALL BOARD"
950 PRINT "4. INPUT TRANSMISSION LOSS VALUES MANUALLY"
960 INPUT "SELECTION: ", S1
970 IF S1 >= 1 AND S1 <= 4 THEN GOTO 980 ELSE GOTO 960
980 PRINT
990 '
1000 ' Statements to determine transmission loss values of wall
1010 '
1020 IF S1 = 3 THEN S7 = 0: S8 = 0: GOSUB 2800
1030 IF S1 = 4 THEN S7 = 0: S8 = 0: GOSUB 2220
1040 IF S1 = 1 THEN S8 = 0: GOSUB 2330
1050 IF S1 = 2 THEN S7 = 0: GOSUB 2560
1060 PRINT
1070 PRINT "DIMENSIONS OF MECH ROOM FLOOR OR WALL ADJACENT TO RECEIVING ROOM"
```

COMPUTER PROGRAMS FOR CHAPTER 7

A6—9

```
1080 INPUT "ENTER LARGER DIMENSION (FT): ", LD
1090 INPUT "ENTER SMALLER DIMENSION (FT): ", SD
1100 '
1110 GOSUB 2980: ' Menu subroutine to determine quality of construction
1120 '
1130 PRINT
1140 '
1150 PRINT "RECEIVER ROOM DIMENSIONS"
1160 INPUT "ENTER RECEIVER ROOM HEIGHT (FT): ", RHR
1170 INPUT "ENTER RECEIVER ROOM LENGTH (FT): ", RLR
1180 INPUT "ENTER RECEIVER ROOM WIDTH  (FT): ", RWR
1190 INPUT "ENTER DISTANCE FROM WALL OR CEILING TO RECEIVER (FT): ", RRR: PRINT
1200 '
1210 PRINT "ACOUSTIC CHARACTERISTICS OF RECEIVER ROOM"
1220 PRINT "1. DEAD:        ACOUSTICAL CEILING, PLUSH CARPET, SOFT FURNISHINGS, PEOPLE"
1230 PRINT "2. MEDIUM DEAD: ACOUSTICAL CEILING, COMMERCIAL CARPET, PEOPLE"
1240 PRINT "3. AVERAGE:     ACOUSTICAL CEILING OR COMMERCIAL CARPET, PEOPLE"
1250 PRINT "4. MEDIUM LIVE: SOME ACOUSTICAL MATERIAL, PEOPLE"
1260 PRINT "5. LIVE:        PEOPLE ONLY"
1270 INPUT "SELECTION: ", S6
1280 IF S6 >= 1 AND S6 <= 5 THEN GOTO 1290 ELSE GOTO 1270
1290 PRINT
1300 PRINT "CONTINUE? (Y OR N)"
1310 Q$ = INKEY$: IF Q$ = "Y" OR Q$ = "N" OR Q$ = "y" OR Q$ = "n" THEN GOTO 1320 ELSE GOTO 1310
1320 IF Q$ = "N" OR Q$ = "n" THEN GOTO 510
1330 '
1340 '
1350 '   Areas and Volumes of source and receiving rooms
1360 '
1370 ' Mechanical equipment room
1380 '
1390 AWM = 2 * (RHM * RLM + RHM * RWM): ' Surface area of walls, mech eq room
1400 ACM = 2 * RWM * RLM: ' Surface area of ceiling and floor, mech eq room
1410 ARM = AWM + ACM: ' Total Surface area of mechanical equipment room
1420 WA = LD * SD: ' Surface area of wall adjacent to receiving room
1430 '
1440 ' Receiving room
1450 '
1460 ARR = 2 * (RHR * RLR + RHR * RWR + RWR * RLR): ' Surface area of receiving room
1470 VOLR = RHR * RLR * RWR: ' Volume of receiving room
1480 '
1490 FOR I = 1 TO 7: AIR(I) = 0: NEXT I: AIR(6) = .0009: AIR(7) = .0029
1500 '
1510 ' Determination of average absorption coefficient
1520 '
1530 FOR I = 1 TO 7
1540 R(I) = (D(1, I) * ACM + D(S2, I) * AWM) / ARM
1550 IF S3 = 1 THEN R(I) = (PCM * D(S4 + 5, I) + (100 - PCM) * R(I)) / 100
1560 IF R(I) > 1 THEN R(I) = 1
1570 NEXT I
1580 '
1590 ' Determination of sound power reduction incident upon wall
1600 '
1610 FOR I = 1 TO 7
1620    TT = WA * ((1 - R(I)) / (ARM * R(I)) + 1 / (4 * WA + 2 * PI * RRM ^ 2))
1630    T1%(I) = 10 * LOG(TT) / LOG(10): 'Sound power reduction incident upon wall
1640 NEXT I
1650 '
1660 T1$ = "CHANGE IN PWL FROM SOURCE TO MECH RM FLOOR/WALL"
1670 '
1680 ' Determination of Sound attenuation through wall
1690 '
1700 IF S1 = 1 THEN FOR I = 1 TO 7: TL(I) = D(13 + S7, I): NEXT I
1710 IF S1 = 2 THEN FOR I = 1 TO 7: TL(I) = D(20 + S8, I): NEXT I
1720 IF S1 = 3 THEN FOR I = 1 TO 7: TL(I) = D(31 + S10, I): NEXT I
1730 '
1740 GOSUB 3100
1750 '
1760 FOR I = 1 TO 7
1770    T2%(I) = -TL(I): ' Transmission loss of common wall
1780 NEXT I
1790 '
1800    T2$ = "TL THROUGH MECH RM FLOOR OR WALL"
1810 '
1820 ' Determination of receiver room sound correction
```

```
1830 '
1840 A = 2 * PI * RRR ^ 2: ' Surface area of hemisphere of radius r
1850 MFP = 4 * VOLR / (ARR * RRR): ' Mean free path/r
1860 FOR I = 1 TO 7
1870     AB = D(S6 + 8, I) + 4 * AIR(I) * VOLR / ARR: RC = ARR * AB / (1 - AB)
1880     ATTN = 10 * LOG(MFP * 4 / RC + 1 / A) / LOG(10) + 10.5
1890     ATTNW = 10 * LOG(4 / RC + 1 / WA) / LOG(10) + 10.5
1900     IF A > WA THEN TT = ATTN ELSE TT = ATTNW
1910     T3%(I) = TT: ' Receiver room sound correction
1920 NEXT I
1930 '
1940 T3$ = "MODIFIED DIFFUSE ROOM CORRECTION"
1950 '
1960 '   Determination of sum of sound power reductions
1970 '
1980 FOR I = 1 TO 7
1990     T%(I) = T1%(I) + T2%(I) + T3%(I)
2000 NEXT I
2010 T$ = "LP(REC RM) - LW(MECH EQUIP RM), DB"
2020 '
2030 '   Output statements
2040 '
2050 PRINT
2060 PRINT SPC(54); "1/1 OCT BAND CENTER FREQ"
2070 PRINT SPC(15); "DESCRIPTION"; SPC(26); " 63  125  250  500   1K   2K   4K"
2080 PRINT STRING$(51, CHR$(45)) + " ---- ---- ---- ---- ---- ---- ----"
2090 PRINT USING "\                                                  \"; T1$;
2100 PRINT SPC(3); USING " ###"; T1%(1); T1%(2); T1%(3); T1%(4); T1%(5); T1%(6); T1%(7)
2110 PRINT USING "\                                                  \"; T2$;
2120 PRINT SPC(3); USING " ###"; T2%(1); T2%(2); T2%(3); T2%(4); T2%(5); T2%(6); T2%(7)
2130 PRINT USING "\                                                  \"; T3$;
2140 PRINT SPC(3); USING " ###"; T3%(1); T3%(2); T3%(3); T3%(4); T3%(5); T3%(6); T3%(7)
2150 PRINT SPC(51); " ---- ---- ---- ---- ---- ---- ----"
2160 PRINT USING "\                                                  \"; T$;
2170 PRINT SPC(3); USING " ###"; T%(1); T%(2); T%(3); T%(4); T%(5); T%(6); T%(7)
2180 END
2190 '
2200 '   Subroutine for manually entering transmission losses through the wall
2210 '
2220 PRINT "ENTER TRANS LOSS VALUES MANUALLY": PRINT
2230 FOR I = 1 TO 7
2240 PRINT TAB(18); "FREQ = ";
2250 PRINT USING "####"; FE(I);
2260 PRINT " HZ: ";
2270 INPUT "TRANS LOSS VALUE (DB) = ", TL(I)
2280 NEXT I
2290 RETURN
2300 '
2310 '   Subroutine for menu of dry-wall configurations
2320 '
2330 PRINT "TRANSMISSION LOSS VALUES OF DRY-WALL CONFIGURATIONS"
2340 PRINT " 1. 3 5/8 IN METAL CHANNEL STUDS; 5/8 IN GYPSUM WALLBOARD SCREWED TO BOTH SIDES"
2350 PRINT "    OF STUDS; ALL JOINTS TAPED AND FINISHED"
2360 PRINT " 2. SAME AS (1) ABOVE EXCEPT 2 IN 3 LB/CU FT DENSITY FIBERGLASS BLANKETS PLACED"
2370 PRINT "    BETWEEN STUDS"
2380 PRINT " 3. 3 5/8 IN METAL CHANNEL STUDS; 2 LAYERS 5/8 IN GYPSUM WALLBOARD SCREWED TO"
2390 PRINT "    BOTH SIDES OF STUDS; ALL EXPOSED JOINTS TAPED AND FINISHED"
2400 PRINT " 4. SAME AS (3) ABOVE EXCEPT 3 IN 3 LB/CU FT DENSITY FIBERGLASS BLANKETS PLACED"
2410 PRINT "    BETWEEN STUDS"
2420 PRINT " 5. 2 1/2 IN METAL CHANNEL STUDS; 2 LAYERS 5/8 IN GYPSUM WALLBOARD SCREWED TO"
2430 PRINT "    BOTH SIDES OF STUDS; ALL EXPOSED JOINTS TAPED AND FINISHED"
2440 PRINT " 6. SAME AS (5) ABOVE EXCEPT 2 IN 3 LB/CU FT DENSITY FIBERGLASS BLANKETS PLACED"
2450 PRINT "    BETWEEN STUDS"
2460 PRINT " 7. 3 5/8 IN METAL CHANNEL STUDS; 2 LAYERS 5/8 IN GYPSUM WALLBOARD SCREWED TO"
2470 PRINT "    HORIZONTAL RESILIENT CHANNELS ON ONE SIDE; 3 LAYERS 5/8 IN GYPSUM WALL-"
2480 PRINT "    BOARD SCREWED TO STUDS ON OTHER SIDE; 3 IN 3 LB/CU FT DNSITY FIBERGLASS"
2490 PRINT "    BETWEEN STUDS"
2500 INPUT "SELECTION: ", S7
2510 IF S7 >= 1 AND S7 <= 7 THEN GOTO 2520 ELSE GOTO 2500
2520 RETURN
2530 '
2540 '   Subroutine for menu of masonry wall/floor/ceiling configurations
2550 '
2560 PRINT "TRANSMISSION LOSS VALUES OF MASONRY WALL/FLOOR/CEILING CONFIGURATIONS"
2570 PRINT " 1. 6X8X18 IN  3-CELL LIGHT CONCRETE MASONRY BLOCK; 21 LB/BLOCK"
```

```
2580 PRINT " 2. 6X8X18 IN  3-CELL DENSE CONCRETE MASONRY BLOCK; 36 LB/BLOCK"
2590 PRINT " 3. 8X8X18 IN  3-CELL LIGHT CONCRETE MASONRY BLOCK; 28 LB/BLOCK"
2600 PRINT " 4. 8X8X18 IN  3-CELL LIGHT CONCRETE MASONRY BLOCK; 34 LB/BLOCK"
2610 PRINT " 5. 8X8X18 IN  3-CELL DENSE CONCRETE MASONRY BLOCK; 48 LB/BLOCK"
2620 PRINT " 6. 3 IN THICK 160 LB/CU FT DENSITY SOLID CONCRETE FLOOR OR WALL;"
2630 PRINT "    SURFACE WEIGHT = 40 LB/SQ FT"
2640 PRINT " 7. 4 IN THICK 144 LB/CU FT DENSITY SOLID CONCRETE FLOOR OR WALL;"
2650 PRINT "    SURFACE WEIGHT = 48 LB/SQ FT"
2660 PRINT " 8. 5 IN THICK 155 LB/CU FT DENSITY SOLID CONCRETE FLOOR OR WALL;"
2670 PRINT "    SURFACE WEIGHT = 65 LB/SQ FT"
2680 PRINT " 9. 6 IN THICK 160 LB/CU FT DENSITY SOLID CONCRETE FLOOR OR WALL;"
2690 PRINT "    SURFACE WEIGHT = 80 LB/SQ FT"
2700 PRINT "10. 8 IN THICK 144 LB/CU FT DENSITY SOLID CONCRETE FLOOR OR WALL;"
2710 PRINT "    SURFACE WEIGHT = 96 LB/SQ FT"
2720 PRINT "11. PREFABRICATED 3 IN DEEP TRAPEZOIDAL CONCRETE CHANNEL SLABS MORTARED"
2730 PRINT "    TOGETHER ON 20 IN CENTERS; SURFACE WEIGHT = 28 LB/SQ FT"
2740 INPUT "SELECTION: ", S8
2750 IF S8 >= 1 AND S8 <= 11 THEN GOTO 2760 ELSE GOTO 2740
2760 RETURN
2770 '
2780 ' Subroutine for menu of painted masonry w & w/o resiliently mounted gypboard
2790 '
2800 PRINT "TRANSMISSION LOSS VALUES OF PAINTED MASONRY BLOCK WALLS AND PAINTED"
2810 PRINT "BLOCK WALLS WITH RESILIENTLY MOUNTED GYPSUM WALL BOARD"
2820 PRINT " 1. 6X8X18 IN 3-CELL LIGHT CONCRETE MASONRY BLOCK; 21 LB/BLOCK"
2830 PRINT "    BOTH SIDES PAINTED WITH PRIMER-SEALER COAT AND COAT OF LATEX"
2840 PRINT " 2. 6X8X18 IN 3-CELL DENSE CONCRETE MASONRY BLOCK; 36 LB/BLOCK"
2850 PRINT "    BOTH SIDES PAINTED WITH PRIMER-SEALER COAT AND COAT OF LATEX"
2860 PRINT " 3. 8X8X18 IN 3-CELL LIGHT CONCRETE MASONRY BLOCK; 28 LB/BLOCK"
2870 PRINT "    BOTH SIDES PAINTED WITH PRIMER-SEALER COAT AND COAT OF LATEX"
2880 PRINT " 4. SAME AS (1) ABOVE EXCEPT 1/2 IN GYPSUM WALLBOARD SCREWED TO"
2890 PRINT "    RESILIENT CHANNELS ON ONE SIDE"
2900 PRINT " 5. SAME AS (3) ABOVE EXCEPT 5/8 IN GYPSUM WALLBOARD SCREWED TO"
2910 PRINT "    RESILIENT CHANNELS ON ONE SIDE"
2920 INPUT "SELECTION: ", S10
2930 IF S10 >= 1 AND S10 <= 5 THEN GOTO 2940 ELSE GOTO 2920
2940 RETURN
2950 '
2960 ' Subroutine for menu of the quality of construnction
2970 '
2980 PRINT
2990 PRINT "QUALITY OF CONSTRUCTION OF MECHANICAL ROOM WALL/FLOOR/CEILING"
3000 PRINT "1. EXCELLENT - NO ACOUSTICAL LEAKS OR PENETRATIONS"
3010 PRINT "2. GOOD - VERY FEW ACOUSTICAL LEAKS AND PENETRATIONS WELL SEALED"
3020 PRINT "3. AVERAGE - MANY ACOUSTICAL LEAKS AND PENETRATIONS POORLY SEALED"
3030 PRINT "4. POOR - MANY ACOUSTICAL LEAKS AND VISIBLE HOLES THROUGH WALL"
3040 INPUT "SELECTION: ", S9
3050 IF S9 >= 1 AND S9 <= 4 THEN GOTO 3060 ELSE GOTO 3040
3060 RETURN
3070 '
3080 ' Subroutine for determining RAT and correcting the transmission loss
3090 '
3100 IF S9 = 1 THEN RAT = .00001: 'EXCELENT CONSTRUCTION
3110 IF S9 = 2 THEN RAT = .0001: 'GOOD CONSTRUCTION
3120 IF S9 = 3 THEN RAT = .001: 'AVG CONSTRUCTION
3130 IF S9 = 4 THEN RAT = .01: 'POOR CONSTRUCTION
3140 FOR I = 1 TO 7
3150 TL(I) = -10 * LOG((1 - RAT) * 10 ^ (-TL(I) / 10) + RAT) / LOG(10)
3160 NEXT I
3170 RETURN
3180 '
3190 ' Sound Absorption associated with construction of walls
3200 '
3210 DATA .01,.01,.01,.02,.02,.02,.03: 'Poured concrete
3220 DATA .11,.1,.05,.06,.07,9.000001E-02,.08: 'Painted concrete block
3230 DATA .17,.18,.22,.15,.15,.19,.12: ' Unpainted concrete block
3240 DATA .31,.29,.1,.06,.05,.05,.05: ' 1 layer 5/8 in Gypsum Board
3250 DATA .13,.1,.05,.05,.05,.05,.05: ' 2 layers 5/8 in Gypsum Board
3260 '
3270 ' Sound Absorption coeficients of acoustic materails
3280 '
3290 DATA .15,.22,.82,1.21,1.1,1.02,1.05: ' 2 in 3 pcf fiberglass
3300 DATA .48,.53,1.19,1.21,1.08,1.01,1.04: ' 3 in 3 pcf fiberglass
3310 DATA .76,.84,1.24,1.24,1.08,1,.97: ' 4 in 3 pcf fiberglass
3320 '
```

```
3330 ' Sound absorption coeffients for use with room constant
3340 '
3350 DATA .26,.30,.35,.4,.43,.46,.52: ' Dead
3360 DATA .24,.22,.18,.25,.3,.36,.42: ' Medium Dead
3370 DATA .25,.23,.17,.2,.24,.29,.34: ' Average
3380 DATA .25,.23,.15,.15,.17,.2,.23: ' Medium Live
3390 DATA .26,.24,.12,.1,0.09,.11,.13: ' Live
3400 '
3410 ' Transmission loss values of dry-wall configuration
3420 '
3430 DATA 11,20,30,37,47,40,44
3440 DATA 14,23,40,45,53,47,48
3450 DATA 19,27,40,46,52,48,48
3460 DATA 24,32,43,50,52,49,50
3470 DATA 16,26,37,45,51,47,49
3480 DATA 22,31,44,50,53,49,51
3490 DATA 27,40,54,61,65,63,67
3500 '
3510 ' Transmission loss values of masonry/floor/ceiling configurations
3520 '
3530 DATA 24,28,34,39,44,50,53
3540 DATA 27,31,36,42,49,50,56
3550 DATA 31,33,37,40,46,51,55
3560 DATA 29,33,39,45,49,56,60
3570 DATA 30,34,40,49,52,59,67
3580 DATA 30,35,39,43,52,58,64
3590 DATA 32,34,35,37,42,49,55
3600 DATA 37,41,43,47,54,59,63
3610 DATA 33,38,43,50,58,64,68
3620 DATA 40,44,47,54,58,63,67
3630 DATA 33,34,34,38,45,55,61
3640 '
3650 ' Transmission loss values of masonry walls with resiliently mounted gypboard
3660 '
3670 DATA 32,36,36,41,45,54,58
3680 DATA 32,36,38,42,49,53,59
3690 DATA 34,38,38,40,46,54,58
3700 DATA 32,36,43,48,61,66,66
3710 DATA 35,39,40,48,59,61,66
```

COMPUTER PROGRAMS FOR CHAPTER 7   A6—13

SAMPLE OUTPUT FOR EXAMPLE 7.3

```
SOUND TRANS THROUGH MECHANICAL EQUIPMENT ROOM FLOOR/CEILING/WALLS

MECHANICAL EQUIPMENT ROOM DIMENSIONS
ENTER MECHANICAL ROOM HEIGHT (FT): 12.75
ENTER MECHANICAL ROOM LENGTH (FT): 12
ENTER MECHANICAL ROOM WIDTH (FT): 14.5
ENTER DISTANCE BETWEEN SOUND SOURCE AND FLOOR OR WALL OF MECH RM (FT): 3

SOUND ABSORPTION ASSOCIATED WITH WALLS
1. POURED CONCRETE
2. CONCRETE BLOCK: PAINTED
3. CONCRETE BLOCK: UNPAINTED
4. SINGLE LAYER 5/8 IN GYPSUM BOARD
5. DOUBLE LAYER 5/8 IN GYPSUM BOARD
SELECTION: 2

ADD ACOUSTICAL ABSORBING MATERIAL TO WALLS AND CEILING? (1. YES    2. NO)
SELECTION: 2

TRANSMISSION LOSS VALUES OF DIFFERENT WALL AND FLOOR CONFIGURATIONS
1. TL VALUES OF DRY-WALL CONFIGURATIONS
2. TL VALUES OF MASONRY WALL/FLOOR/CEILING CONFIGURATIONS
3. TL VALUES OF PAINTED MASONRY BLOCK WALLS AND PAINTED
   BLOCK WALLS WITH RESILIENTLY MOUNTED GYPSUM WALL BOARD
4. INPUT TRANSMISSION LOSS VALUES MANUALLY
SELECTION: 3

TRANSMISSION LOSS VALUES OF PAINTED MASONRY BLOCK WALLS AND PAINTED
BLOCK WALLS WITH RESILIENTLY MOUNTED GYPSUM WALL BOARD

1. 6X8X18 IN 3-CELL LIGHT CONCRETE MASONRY BLOCK; 21 LB/BLOCK
   BOTH SIDES PAINED WITH PRIMER-SEALER COAT AND COAT OF LATEX
2. 6X8X18 IN 3-CELL DENSE CONCRETE MASONRY BLOCK; 36 LB/BLOCK
   BOTH SIDES PAINED WITH PRIMER-SEALER COAT AND COAT OF LATEX
3. 8X8X18 IN 3-CELL LIGHT CONCRETE MASONRY BLOCK; 28 LB/BLOCK
   BOTH SIDES PAINED WITH PRIMER-SEALER COAT AND COAT OF LATEX
4. SAME AS (1) ABOVE EXCEPT 1/2 IN GYPSUM WALLBOARD SCREWED TO
   RESILIENT CHANNELS ON ONE SIDE
5. SAME AS (3) ABOVE EXCEPT 5/8 IN GYPSUM WALLBOARD SCREWED TO
   RESILIENT CHANNELS ON ONE SIDE
SELECTION: 3

DIMENSIONS OF MECH ROOM FLOOR OR WALL ADJACENT TO RECEIVING ROOM
ENTER LARGER DIMENSION (FT): 14.5
ENTER SMALLER DIMENSION (FT): 9

QUALITY OF CONSTRUCTION OF MECHANICAL ROOM WALL/FLOOR/CEILING
1. EXCELLENT - NO ACOUSTICAL LEAKS OR PENETRATIONS
2. GOOD - VERY FEW ACOUSTICAL LEAKS AND PENETRATIONS WELL SEALED
3. AVERAGE - MANY ACOUSTICAL LEAKS AND PENETRATIONS POORLY SEALED
4. POOR - MANY ACOUSTICAL LEAKS AND VISIBLE HOLES THROUGH WALL
SELECTION: 1

RECEIVER ROOM DIMENSIONS:
ENTER RECEIVER ROOM HEIGHT (FT): 9
ENTER RECEIVER ROOM LENGTH (FT): 12
ENTER RECEIVER ROOM WIDTH (FT): 14.5
ENTER DISTANCE FROM WALL OR CEILING TO RECEIVER (FT): 6

ACOUSTIC CHARACTERISTICS OF RECEIVER ROOM
1. DEAD:        ACOUSTICAL CEILING, PLUSH CARPET, SOFT FURNISHINGS, PEOPLE
2. MEDIUM DEAD: ACOUSTICAL CEILING, COMMERCIAL CARPET, PEOPLE
3. AVERAGE:     ACOUSTICAL CEILING OR COMMERCIAL CARPET, PEOPLE
4. MEDIUM LIVE: SOME ACOUSTICAL MATERIAL, PEOPLE
5. LIVE:    PEOPLE ONLY
SELECTION: 2

DO YOU WANT TO CONTINUE? (Y OR N): Y
```

DESCRIPTION	\multicolumn{7}{c}{1/1 OCT BAND CENTER FREQ}
	63	125	250	500	1K	2K	4K
CHANGE FROM SOURCE TO MECH RM	2	3	6	5	4	3	3
TL THROUGH MECH RM FLOOR OR WALL	−34	−38	−38	−40	−45	−49	−49
MODIFIED DIFFUSE ROOM CORRECTION	−6	−5	−4	−6	−7	−8	−9
LP(REC RM) − LW(MECH EQUIP RM), DB	−38	−40	−36	−41	−48	−54	−55

COMPUTER PROGRAMS FOR CHAPTER 7

PROGRAM OUTLINE

Input Data from Table 7.5
Input system parameters:
    Mechanical room dimensions
    Distance between sound source and hole in wall
    Sound absorption associated with construction of walls
    Type and amount of acoustical absorbing material added
Calculate:
    Surface areas of mechanical room
    Average absorption coefficient of mechanical room
    Change in sound power level from sound source to hole in wall, dB

COMPUTER PROGRAM – MECHRC.BAS

```
10  ' >TRANSMISSION OF SOUND TRHOUGH HOLE IN MECHANICAL ROOM WALL COMPUTER PROGRAM<
20  DIM T%(7), D(8, 7), EF(50), L$(8)
30  PI = 3.141593
40  '
50  '    LIST OF VARIABLES
60  '
70  '    ACM = surface area of mechanical room floor and ceiling, sq. ft.
80  '    ARM = surface area of mechanical room, sq. ft.
90  '    AWM = surface area of mechanical room walls, sq. ft.
100 '    D() = absorption coefficients
110 '    LD = larger dimension of opening in mechanical room wall, in
120 '    PCM = percent surface area covered by acoustical material, %
130 '    R() = 1/1 octave band average absorption coefficient
140 '    RHM = mechanical room height, ft
150 '    RLM = mechanical room length, ft
160 '    RRM = distance between sound source and hole in mech. eq. room wall, ft
170 '    RWM = mechanical room width, ft
180 '    SD = smaller dimension of opening in mechanical room wall, in
190 '    S1 = selection for construction of walls
200 '    S2 = selection to determine whether to add acoustical absorbing material
210 '    S3 = selection for thickness of insulation
220 '    AR = Used in equation 7.13
230 '    TT = Used in equation 7.13
240 '    T%() = Sound power transmission loss, dB (eq. 7.13)
250 '    WA = area of hole in mechanical room wall, sq. ft.
260 '
270 '
280 FOR I = 1 TO 8
290   FOR J = 1 TO 7
300     READ D(I, J): ' Reads in absorption coefficients
310   NEXT J
320 NEXT I
330 '
340 CLS : PRINT "CHANGE IN SOUND POWER LEVELS"
350 PRINT "FROM SOUND SOURCE TO HOLE IN MECHANICAL EQUIPMENT ROOM WALL": PRINT
360 PRINT "MECHANICAL EQUIPMENT ROOM"
370 INPUT "ENTER MECHANICAL ROOM HEIGHT (FT): ", RHM
380 INPUT "ENTER MECHANICAL ROOM LENGTH (FT): ", RLM
390 INPUT "ENTER MECHANICAL ROOM WIDTH (FT): ", RWM
400 INPUT "ENTER DISTANCE BETWEEN SOUND SOURCE AND HOLE IN MECH EQUIP RM WALL (FT): ", RRM
410 '
420 PRINT : PRINT "SOUND ABSORPTION ASSOCIATED WITH CONSTRUCTION OF WALLS"
430 PRINT "1. POURED CONCRETE"
440 PRINT "2. CONCRETE BLOCK: PAINTED"
450 PRINT "3. CONCRETE BLOCK: UNPAINTED"
460 PRINT "4. SINGLE LAYER 5/8 IN GYPSUM BOARD"
470 PRINT "5. DOUBLE LAYER 5/8 IN GYPSUM BOARD"
480 '
490 INPUT "SELECTION: ", S1
500 IF S1 >= 1 AND S1 <= 5 THEN GOTO 510 ELSE GOTO 490
510 PRINT
520 INPUT "ADD ACOUSTICAL ABSORBING MATERIAL TO WALLS AND CEILING?  (1. YES    2. NO): ", S2
530 IF S2 >= 1 AND S2 <= 2 THEN GOTO 540 ELSE GOTO 520
```

```
540 IF S2 = 2 THEN S3 = 0: PCM = 0: GOTO 620
550 PRINT
560 PRINT "1. 2 IN 3 LB/CU FT FIBERGLASS INSULATION"
570 PRINT "2. 3 IN 3 LB/CU FT FIBERGLASS INSULATION"
580 PRINT "3. 4 IN 3 LB/CU FT FIBERGLASS INSULATION"
590 INPUT "SELECTION: ", S3
600 IF S3 >= 1 AND S3 <= 3 THEN GOTO 610 ELSE GOTO 590
610 INPUT "ENTER PERCENT WALLS, FLOOR AND CEILING COVERED BY ACOUSTICAL MATERIAL: ", PCM
620 PRINT " ": PRINT "DIMENSIONS OF OPENING IN MECHANICAL ROOM WALL"
630 INPUT "ENTER LARGER DIMENSION (IN): ", LD
640 INPUT "ENTER SMALLER DIMENSION (IN): ", SD
650 PRINT
660 PRINT "CONTINUE? (Y OR N)"
670 Q$ = INKEY$: IF Q$ = "Y" OR Q$ = "N" OR Q$ = "y" OR Q$ = "n" THEN GOTO 680 ELSE GOTO 670
680 IF Q$ = "N" OR Q$ = "n" THEN GOTO 340
690 LD = LD / 12: SD = SD / 12
700 '
710 AWM = 2 * (RHM * RLM + RHM * RWM): ' Surface area of mech. rm. walls
720 ACM = 2 * (RWM * RLM): ' Surface area of mechanical room floor and ceiling
730 ARM = AWM + ACM: ' Surface area of mechanical room
740 WA = LD * SD: ' area of hole in mechanical room wall
750 '
760 FOR I = 1 TO 7
770 '    Average absorption coefficient
780    R(I) = (D(1, I) * ACM + D(S1, I) * AWM) / ARM: ' Equation 7.14
790    IF S2 = 1 THEN R(I) = (PCM * D(S3 + 5, I) + (100 - PCM) * R(I)) / 100: ' Eq. 7.15
800    IF R(I) > 1 THEN R(I) = 1
810 NEXT I
820 AR = 1 / (2 * PI * RRM ^ 2): ' For use in equation 7.13
830 FOR I = 1 TO 7
840    TT = (1 - R(I)) / (ARM * R(I)) + AR: ' for use in equation 7.13
850    T%(I) = 10 * LOG(TT) / LOG(10) + 10 * LOG(WA) / LOG(10): ' Equation 7.13
860 NEXT I
870 '
880 ' Output statements
890 '
900 PRINT
910 T1$ = "CHANGE IN PWL FROM SOURCE TO MECH RM WALL HOLE"
920 PRINT SPC(54); "1/1 OCT BAND CENTER FREQ"
930 PRINT SPC(15); "DESCRIPTION"; SPC(26); " 63  125  250  500  1K   2K   4K"
940 PRINT STRING$(51, CHR$(45)) + " ___ ___ ___ ___ ___ ___ ___"
950 PRINT USING "\                                                 \"; T1$;
960 PRINT SPC(3); USING " ###"; T%(1); T%(2); T%(3); T%(4); T%(5); T%(6); T%(7)
970 END
980 '
990 '    Sound absorption coefficients
1000 '
1010 DATA 0.01,0.01,0.01,0.02,0.02,0.02,0.03: ' POURED CONCRETE
1020 DATA 0.11,0.10,0.05,0.06,0.07,9.000001E-02,0.08: ' CONCRETE BLOCK: Painted
1030 DATA 0.17,0.18,0.22,0.15,0.15,0.19,0.12: ' CONCRETE BLOCK: Unpainted
1040 DATA 0.31,0.29,0.10,0.06,0.05,0.05,0.05: ' SINGLE LAYER 5/8 IN GYPSUM BOARD
1050 DATA 0.13,0.10,0.05,0.05,0.05,0.05,0.05: ' DOUBLE LAYER 5/8 IN GYPSUM BOARD
1060 DATA 0.15,0.22,0.82,1.21,1.10,1.02,1.05: ' 2 IN 3 LB/CU FT FIBERGLASS INS.
1070 DATA 0.48,0.53,1.19,1.21,1.08,1.01,1.04: ' 3 IN 3 LB/CU FT FIBERGLASS INS.
1080 DATA 0.76,0.84,1.24,1.24,1.08,1.00,0.97: ' 4 IN 3 LB/CU FT FIBERGLASS INS.
```

SAMPLE OUTPUT EXAMPLE 7.4

```
CHANGE IN SOUND POWER LEVELS
FROM SOUND SOURCE TO HOLE IN MECHANICAL EQUIPMENT ROOM WALL

MECHANICAL EQUIPMENT ROOM
ENTER MECHANICAL ROOM HEIGHT (FT): 12.75
ENTER MECHANICAL ROOM LENGTH (FT): 12
ENTER MECHANICAL ROOM WIDTH (FT): 14.5
ENTER DISTANCE BETWEEN SOUND SOURCE AND HOLE IN MECH EQUIP RM WALL (FT): 6

SOUND ABSORPTION ASSOCIATED WITH CONSTRUCTION OF WALLS
1. POURED CONCRETE
2. CONCRETE BLOCK: PAINTED
3. CONCRETE BLOCK: UNPAINTED
4. SINGLE LAYER 5/8 IN GYPSUM BOARD
5. DOUBLE LAYER 5/8 IN GYPSUM BOARD
SELECTION: 2

ADD ACOUSTICAL ABSORBING MATERIAL TO WALLS AND CEILING?   (1. YES 2. NO): 2

DIMENSIONS OF OPENING IN MECHANICAL ROOM WALL
ENTER LARGER DIMENSION (IN): 48
ENTER SMALLER DIMENSION (IN): 36

CONTINUE? (Y OR N)
```

DESCRIPTION	63	125	250	500	1K	2K	4K
CHANGE FROM SOURCE TO MECH RM WALL HOLE	−7	−7	−4	−5	−6	−7	−6

1/1 OCT BAND CENTER FREQ

A6–18　APPENDIX 6

SOUND ATTENUATION IN OUTDOOR ENVIRONMENTS

PROGRAM OUTLINE

Input regression coefficients
Input system parameters:
    1/1 Octave band sound power levels, dB
    Distance from sound source to receiver, ft
    Distances from ground to sound source and receiver
    Barrier information
    Reflecting Surface information
Calculate:
    Sound Pressure Level at source before corrections
    Barrier attenuation if a barrier is present
    Correction for vertical reflecting surface if present
    Final sound pressure level after corrections

COMPUTER PROGRAM – OUTDOOR.BAS

```
10  '    >Outdoor Sound Attenuation<
20  '
30  '      List of Variables
40  '
50  '     A = Distance from sound source to barrier
60  '     AA = Equation 7.28
70  '     ATTN() = barrier attenuation, dB
80  '     B = Distance from wall to receiver, ft
90  '     BB = Equation 7.29
100 '     DISCOR() = Distance correction, dB
110 '     DD = Equation 7.30
120 '     CO = speed of sound, ft/s
130 '     DL = correction term for a reflecting surface
140 '     FZ = 1/1 Octave band Center frequency
150 '     H = Height of barrier, ft
160 '     HR = Distance from receiver to ground, ft
170 '     HS = Distance from sound source to ground, ft
180 '     L = Horizontal distance between sound source and receiver, ft
190 '     RI = Distance from image source to receiver, ft
200 '     RS = Distance from sound source to receiver, ft
210 '     S = Selection to determine whether or not there is a barrier
220 '     S1 = Selection for proximaty to a reflecting surface
230 '     TT() = Original Sound Power Level, dB
240 '     T%() = Sound Pressure Level at receiver, dB
250 '
260 DIM TT(8), T%(8), DISCOR(8)
270 DIM LAMBDA(8), ATTN(8), N(8)
280 PI = 3.141593
290 CLS
300 FOR I = 1 TO 8
310     FZ = 62.5 * 2 ^ (I - 1)
320     PRINT "ENTER "; FZ; " HZ 1/1 OCTAVE BAND SOUND POWER LEVEL"; : INPUT TT(I)
330 NEXT I
340 PRINT "CONTINUE? (Y OR N)"
350 Q$ = INKEY$: IF Q$ = "Y" OR Q$ = "N" OR Q$ = "y" OR Q$ = "n" THEN GOTO 360 ELSE GOTO 350
360 IF Q$ = "N" OR Q$ = "n" THEN CLS : GOTO 300
370 PRINT
380 INPUT "DISTANCE FROM SOUND SOURCE TO GROUND, FT: ", HS
390 INPUT "DISTANCE FROM RECEIVER TO GROUND, FT: ", HR
400 INPUT "HORIZONTAL DISTANCE FROM SOUND SOURCE TO RECEIVER, FT: ", L
410 PRINT
420 INPUT "IS SOUND SOURCE NEAR A VERTICAL HARD REFLECTING SURFACE? (1. YES    2. NO): ", S1
430 IF S1 = 1 THEN INPUT "DISTANCE BETWEEN VERTICAL WALL AND SOURCE, FT: ", DW
440 PRINT
450 INPUT "IS THERE A BARRIER? (1. YES    2. NO): ", S
460 IF S = 2 THEN GOTO 500
470 INPUT "DISTANCE FROM SOUND SOURCE TO BARRIER, FT: ", A
```

```
480 B = RS - A
490 INPUT "HEIGHT OF BARRIER, FT: ", H
500 PRINT
510 PRINT "CONTINUE? (Y OR N)"
520 Q$ = INKEY$: IF Q$ = "Y" OR Q$ = "N" OR Q$ = "y" OR Q$ = "n" THEN GOTO 530 ELSE GOTO 520
530 IF Q$ = "N" OR Q$ = "n" THEN CLS : GOTO 370
540 '
550 IF S = 1 THEN GOSUB 930
560 GOSUB 790
570 IF S = 1 THEN FOR I = 1 TO 8: T%(I) = T%(I) + ATTN(I): NEXT I
580 ' Output statements
590 '
600 PRINT
610 PRINT SPC(45); "1/1 OCT BAND CENTER FREQ"
620 PRINT SPC(10); "DESCRIPTION"; SPC(21); " 63 125 250 500 1K 2K 4K 8K"
630 PRINT STRING$(41, CHR$(45)) + " ___ ___ ___ ___ ___ ___ ___ ___"
640 PRINT USING "\                                          \"; "ORIGINAL SOUND POWER LEVEL, dB";
650 PRINT USING " ###"; TT(1); TT(2); TT(3); TT(4); TT(5); TT(6); TT(7); TT(8)
660 PRINT USING "\                                          \"; "DISTANCE CORRECTION, dB";
670 PRINT USING " ###"; DISCOR(1); DISCOR(2); DISCOR(3); DISCOR(4); DISCOR(5); DISCOR(6); DISCOR(7); DISCOR(8)
680 IF S = 1 THEN PRINT USING "\                                          \"; "BARRIER ATTEN., dB";
690 IF S = 1 THEN PRINT USING " ###"; ATTN(1); ATTN(2); ATTN(3); ATTN(4); ATTN(5); ATTN(6); ATTN(7); ATTN(8)
700 IF S1 = 1 THEN PRINT USING "\                                          \"; "CORR. FACTOR FOR A REFLECTING WALL, dB";
710 IF S1 = 1 THEN PRINT USING " ###"; DL; DL; DL; DL; DL; DL; DL; DL
720 PRINT USING "\                                          \"; "CORRECTION FOR HALF SPACE, dB";
730 PRINT USING " ###"; 3; 3; 3; 3; 3; 3; 3; 3
740 PRINT STRING$(41, CHR$(45)) + " ___ ___ ___ ___ ___ ___ ___ ___"
750 PRINT USING "\                                          \"; T$;
760 PRINT USING " ###"; T%(1); T%(2); T%(3); T%(4); T%(5); T%(6); T%(7); T%(8)
770 END
780 '
790 DL = 0
800 RS = SQR(L ^ 2 + (HR - HS) ^ 2)
810 RI = SQR((L + 2 * DW) ^ 2 + (HR - HS) ^ 2)
820 LOGR = LOG(RI / RS) / LOG(10)
830 ' Correction for case when sound source is close to a reflecting surface
840 IF S1 = 1 THEN DL = 3.0046 - 9.29459 * LOGR + 10.127 * LOGR ^ 2 - 3.8424 * LOGR ^ 3
850 RREF = .9252: ' Reference in ft.
860 FOR I = 1 TO 8
870     DISCOR(I) = -20 * LOG(RS / RREF) / LOG(10)
880     T%(I) = TT(I) + DISCOR(I) + 3 + DL
890 NEXT I
900 T$ = "SOUND PRESS. LEVEL AT RECEIVER, dB"
910 RETURN
920 '
930 ' Subroutine for barriers
940 '
950 C0 = 1125.3: ' Speed of sound (ft/s)
960 '
970 B = L - A
980 AA = SQR(A ^ 2 + (H - HS) ^ 2)
990 BB = SQR(B ^ 2 + (H - HR) ^ 2)
1000 DD = SQR((A + B) ^ 2 + (HS - HR) ^ 2)
1010 PRINT
1020 FOR I = 1 TO 8
1030     FZ = 62.5 * 2 ^ (I - 1)
1040     LAMBDA = C0 / FZ
1050     N = (2 / LAMBDA) * (AA + BB - DD)
1060     LOGN = LOG(N) / LOG(10)
1070     ATTN(I) = -(13.2383 + 7.7983 * LOGN + 1.1727 * LOGN ^ 2 - .37422 * LOGN ^ 3)
1080     IF ATTN(I) < -24 THEN ATTN(I) = -24
1090 NEXT I
1100 RETURN
```

SAMPLE OUTPUT FOR EXAMPLE 7.5

```
ENTER    62.5   HZ 1/1 OCTAVE BAND SOUND POWER LEVEL? 91
ENTER    125    HZ 1/1 OCTAVE BAND SOUND POWER LEVEL? 87
ENTER    250    HZ 1/1 OCTAVE BAND SOUND POWER LEVEL? 83
ENTER    500    HZ 1/1 OCTAVE BAND SOUND POWER LEVEL? 82
ENTER    1000   HZ 1/1 OCTAVE BAND SOUND POWER LEVEL? 78
ENTER    2000   HZ 1/1 OCTAVE BAND SOUND POWER LEVEL? 76
ENTER    4000   HZ 1/1 OCTAVE BAND SOUND POWER LEVEL? 72
ENTER    8000   HZ 1/1 OCTAVE BAND SOUND POWER LEVEL? 69
```

DISTANCE FROM SOUND SOURCE TO GROUND, FT: 2
DISTANCE FROM RECEIVER TO GROUND, FT: 5
HORIZONTAL DISTANCE FROM SOUND SOURCE TO RECEIVER, FT: 15

IS SOUND SOURCE NEAR A VERTICAL HARD REFLECTING SURFACE? (1. YES 2. NO): 1
DISTANCE BETWEEN VERTICAL WALL AND SOURCE, FT: 3

IS THERE A BARRIER? (1. YES 2. NO): 1
DISTANCE FROM SOUND SOURCE TO BARRIER, FT: 5
HEIGHT OF BARRIER, FT: 6

CONTINUE? (Y OR N)

DESCRIPTION	\multicolumn{8}{c}{1/1 OCT BAND CENTER FREQ}
	63	125	250	500	1K	2K	4K	8K
ORIGINAL SOUND POWER LEVEL, dB	91	87	83	82	78	76	72	69
DISTANCE CORRECTION, dB	−24	−24	−24	−24	−24	−24	−24	−24
BARRIER ATTEN., dB	−7	−9	−11	−13	−16	−18	−21	−24
CORR. FACTOR FOR A REFL. WALL, dB	2	2	2	2	2	2	2	2
CORRECTION FOR HALF SPACE, dB	3	3	3	3	3	3	3	3
SOUND PRESS. LEVEL AT RECEIVER, dB	65	59	53	50	43	39	32	26